THERE IS NO PLANET B

Updated Edition

Feeding the world, climate change, biodiversity, antibiotics, plastics – the list of concerns seems endless. Given the global nature of the challenges we face, what can any of us DO?

Mike Berners-Lee has crunched the numbers and plotted a course of action that is practical and enjoyable. Readers will find the big-picture perspective on the environmental and economic challenges of the day laid out in one place, and traced through to the underlying roots – questions of how we live and think on Planet A.

This updated edition includes an expanded 'What can I do?' section; a new chapter on protest, including thoughts on Extinction Rebellion and school children marches; more on offsetting, carbon net zero, and investing, in response to developments in the business world; and more on pandemics, COVID-19 and wildfires.

Mike Berners-Lee thinks, writes, researches and consults on sustainability and responses to the challenges of the twenty-first century. He is the founder of Small World Consulting (SWC), an associate company of Lancaster University, which works with organisations from small businesses to the biggest tech giants. SWC is a leader in the field of carbon metrics, targets and actions. About his first book – *How Bad Are Bananas? The Carbon Footprint of Everything* – Bill Bryson wrote 'I can't remember the last time I read a book that was more fascinating, useful and enjoyable all at the same time.' His second book (co-written with Duncan Clark) – *The Burning Question: We Can't Burn Half the World's Oil, Coal, and Gas. So How Do We Quit?* – explores the big picture of climate change and the underlying global dynamics, asking what mix of politics, economics, psychology and technology are really required to deal with the problem. Al Gore described it as ‘Fascinating, important and highly recommended.’ Mike is a professor in the Institute for Social Futures at Lancaster University, where he develops practical tools for thinking about the future, and researches the global food system and carbon metrics.

'. . . a sort of Alexa to tell you how to live in a more planet-friendly fashion . . . Amazingly, it manages to make the complexities of planet-scale economic and environmental interconnectivity fun: a platter of potential doom, served with a smiley face and sparkler . . . There Is No Planet B *is a rallying cry for a generation worried that they will inherit a world shorn of nature's wonders and of the freedoms and opportunities we take for granted. Buying the book and adopting its key guidelines and mindset will go a long way to ensuring the planet we hand on may just be liveable.'*

Adrian Barnett, *New Scientist*

'. . . a "handbook" for how humanity can thrive in dark times . . . how can we keep living happily on this planet?'

Leslie Hook, *Financial Times*

'A comforting handbook for anyone frozen into inaction.'

Sunday Times

'Who should read There is No Planet B? ***Everyone****. Mike Berners-Lee has written a far-ranging and truth-telling handbook that is as readable as it is instructive.'*

Elizabeth Kolbert, *The New Yorker* and author of *The Sixth Extinction: An Unnatural History*

'. . . a lively and cogent assessment of what is happening to the Earth's biosphere and resources . . . tells us what we can do if we want to make a difference, and tread more softly on the planet. All citizens should be grateful for this information-packed and wide-ranging primer.'

Martin Rees, Astronomer Royal

'This is a massively entertaining compendium of bite-sized facts . . . It's also massively important, given the current state of the planet.'

Bill McKibben

'... an unexpected aesthetic pleasure as well as a guide to action. No matter how much you already know, this book will help orient you to where we are now on this, the only planet we have ... It would be best if everyone read it.'

Kim Stanley Robinson, author of the *Mars* trilogy and *New York 2140*

'... a wonderfully abundant buffet-table of knowledge about sustainability and you can enjoy it all at one sitting or benefit from visiting for bite-size chunks. Either way, you'll come away wiser, healthier and also entertained ... [Mike] shares his insights with warmth and wit, and his book could not be more timely.'

David Shukman, BBC Science Editor

'I absolutely love this book. Evidence-based, robust, and full of practical guidance. In an increasingly complex and confusing world, this book stands out as a beacon of common sense, clarity and v crucially – hope.'

Caroline Lucas, Member of Parliament, United Kingdom

There Is No Planet B

A Handbook for the Make or Break Years

Updated Edition

MIKE BERNERS-LEE

CAMBRIDGE
UNIVERSITY PRESS

University Printing House, Cambridge CB2 8BS, United Kingdom

One Liberty Plaza, 20th Floor, New York, NY 10006, USA

477 Williamstown Road, Port Melbourne, VIC 3207, Australia

314–321, 3rd Floor, Plot 3, Splendor Forum, Jasola District Centre, New Delhi – 110025, India

79 Anson Road, #06–04/06, Singapore 079906

Cambridge University Press is part of the University of Cambridge.

It furthers the University's mission by disseminating knowledge in the pursuit of education, learning, and research at the highest international levels of excellence.

www.cambridge.org
Information on this title: www.cambridge.org/9781108821575
DOI: 10.1017/9781108900997

First published 2021

Printed in the United Kingdom by TJ Books Ltd, Padstow Cornwall

A catalogue record for this publication is available from the British Library.

Library of Congress Cataloging-in-Publication Data
Names: Berners-Lee, Mike, 1964– author.
Title: There is no Planet B : a handbook for the make or break years / Mike Berners-Lee.
Description: Updated edition. | New York : Cambridge University Press, 2021. | Includes bibliographical references and index.
Identifiers: LCCN 2020025561 (print) | LCCN 2020025562 (ebook) | ISBN 9781108821575 (paperback) | ISBN 9781108900997 (epub)
Subjects: LCSH: Power resources. | Power resources–Environmental aspects. | Energy consumption. | Energy consumption–Environmental aspects. | Climatic changes. | Pollution. | Environmental protection.
Classification: LCC TJ163.2 .B4745 2020 (print) | LCC TJ163.2 (ebook) | DDC 363.7–dc23
LC record available at https://lccn.loc.gov/2020025561
LC ebook record available at https://lccn.loc.gov/2020025562

ISBN 978-1-108-82157-5 Paperback

How is this book laid out?

Almost the whole book is in the form of questions. This means you can read it in different ways: you can dip in at random, or search for topics by scanning the contents or index, or read it from end to end in what I hope is a logical flow.

The first few sections deal with obvious physical, technical and scientific challenges, moving towards deeper underlying issues, and then into the terrain of values, truth and finally a discussion of ways we must learn to think in order to cope with our new era.

Towards the back I have included an **Alphabetical Quick Tour**. This is supposed to be both fun and useful. It also gives me a chance to include a few things that don't fit properly anywhere else but deserve a mention. Presenting things alphabetically creates a totally new order with a random logic. I hope this helps to reinforce the idea that although most of the book is told in sequence, it all interrelates so much that we need to hold it all in our heads at the same time.

The endnotes are intended to be read if you want a bit more detail. They are not just references. Often there is good stuff in there that I have kept out of the main text simply to help the flow.

Finally, a note on language. I've kept it as simple and jargon-free as possible, because I hope this book will be read, enjoyed and used by a very wide audience.

To the memory of mum

CONTENTS

Acknowledgements page xvi
What's New in this Updated Edition? xvii

INTRODUCTION TO THE FIRST EDITION 1
Welcome to a new era 1
A handbook of everything 3
When it's all so global, what can I do? 5
What values underpin this book? 7
What can we aim for? 9
Not the last word . . . 10

1 FOOD 12
How much food energy do we need to eat? 13
How much food do we grow worldwide? 13
What happens to the food we grow? 13
Given the global surplus, why are some people malnourished? 16
Why don't more people explode from overeating? 17
How many calories do we get from animals? 18
How much do animals help with our protein supply? 18
Do we need animals for iron, zinc or vitamin A? 20
How much of our antibiotics are given to animals? 21
Do factory farms make pandemics more likely? 22
How much deforestation do soya beans cause? 24
What's the carbon footprint of agriculture? 25
What are the carbon footprints of different foods? 26
Should I go veggie or vegan? 30
What can shops do about meat and dairy habits? 31
What can restaurants do? 32
What can farmers and governments do? 32

How could one crop save us over half a billion tonnes CO_2e? 33
Is local food best? 34
Where does fish fit in? 36
When is a seabass not a seabass? 37
How can we sustain our fish? 38
What food is wasted, where and how? 40
How can we cut the world's waste? 45
Why don't supermarkets care more about their waste? 47
When food can't be sold or eaten, what should be done with it? 47
How much food goes to biofuel? 49
How many farmers does the world need? 50
How can new technologies help feed the world? 50
How can we produce enough food for 9.7 billion of us in 2050? 52
Why do we all need to know our food supply chains? 54
What investments are needed into food land and sea? 55
Food action summary: What can I do and what can be done? 56

2 MORE ON CLIMATE AND ENVIRONMENT 58
What are the 14 things that every politician needs to know about the climate emergency? 58
What are the biodiversity stats? And why do they matter? 60
What is ocean acidification and why does it matter? 62
How much plastic is there in the world? 63
Is fossil fuel better burned or turned into plastic? 65

3 ENERGY 66
How much do we use? 66
How has our use changed over time? 67
What do we use it *for*? 69
Where do we get it all from? 70
How bad are fossil fuels? 72

How much energy comes from the sun? 74
Can the sun's energy be harnessed? 74
How much solar power could we *ever* have? 75
Which countries have the most sunlight? 77
Which countries have the least sun per person? 79
What about when the sun isn't shining? 80
How useful is wind energy? 82
Which countries have the most wind per person? 83
Why is sun better than rain? 83
Is nuclear nasty? 85
Would fusion solve everything? 87
Are biofuels bonkers? 88
Should we frack? 89
Does more renewables mean less fossil fuel? 91
What is the catch with energy efficiency? 92
Given the catch, what can efficiency do for us? 94
Why is cleaning our electricity just the easy part of the transition from fossil fuels? 95
How can we keep the fuel in the ground? 97
Who has the most fossil fuel and how will they cope? 99
Will we need to take carbon back out of the air? 102
Can carbon be offset? 105
How much energy are we on track to use in 2100? 108
Can enough energy ever be enough? 110
Energy solution summary 111
Energy: What can I do? 112

4 TRAVEL AND TRANSPORT 114
How much do we travel today? 114
How much travel will we want in the future? 116
How many travel miles can we get from a square metre of land? 117
How can we sort out urban transport? 119
Will shared transport make life better or worse? 120
Should I buy an electric car? 122

How urgently should I ditch my diesel? 123
Could autonomous cars be a disaster? Or brilliant? 126
How can we fly in the low carbon world? 126
Should I fly? 129
Do virtual meetings save energy and carbon? 129
How bad are boats? And can they be electrified? 130
E-bikes or pedals? 132
When might we emigrate to another planet? 133

5 GROWTH, MONEY AND METRICS 136
Which kinds of growth can be healthy in the Anthropocene? 137
Why is GDP such an inadequate metric? 141
How do our metrics need to change? 142
What metrics do we need to take more note of? 143
What metrics do we need to downgrade? 144
Can the free market deal with Anthropocene challenges? 145
Which is better, the market economy or the planned economy? 146
What is trickledown and why is it dangerous? 147
Why might wealth distribution matter more than ever? 149
How is the world's wealth distributed? 149
Why are most Americans so much poorer than most Italians? 151
How has wealth distribution been changing? 153
When is wealth distributed like the energy in a gas? (And when is it not?) 154
How can human wealth become more like the energy in a gas? 156
What should we invest in? 159
How can these essential investments be funded? 160
What can fund managers do? 161
Why does the right tax make us better off? 162
Do we need a carbon price? 166

How expensive will carbon need to become? 167
How should I spend my money? 168

6 PEOPLE AND WORK 169
Does it all come down to population? 169
What can I do to help with population? 170
When is a 'job' a good thing? 171
How much of a person should come to work? 173
Why would anyone work if they already had a citizen's wage? 174
What are my chances of being in prison? 175

7 BUSINESS AND TECHNOLOGY 179
When is it good that an organisation exists? 179
How can businesses think about the world? 180
How can a business think systemically? 181
What is a science-based target? 185
What is so special when science-based targets are applied to the supply chain? 186
Do we drive technology growth, or does it drive us? 188
How can we take control of technology? 189

8 VALUES, TRUTH AND TRUST 191
What is the evidence base to choose some values over others? 191
What values do we need to be the new global cultural norms? 193
Can we deliberately change our values? 194
What makes our values change? 195
Is there even such a thing as 'truth' or 'facts'? 198
Is 'truth' personal? 199
Why is dedication to truth more important than ever? 199
What is a culture of truth? 200
Is it possible to have a more truthful culture? 200
What can *I* do to promote a culture of truth? 201

What can journalists do to promote truth? 201
What can politicians do? 202
How can I work out who and what to trust? 203
What are some bad reasons for placing trust? 204
How can I tell whether to trust anything in this book? 205

9 THINKING SKILLS FOR TODAY'S WORLD 207
What new ways of thinking do we need in the twenty-first century? 207
How can twenty-first century thinking skills be developed? 213
Where is religion and spirituality in all this? 214

10 PROTEST 217
Do we need to protest? 217
What has been Extinction Rebellion's magic? 218
What is the next evolution of protest? 219
Should children protest? 220

11 BIG-PICTURE SUMMARY 222
Rising human power has taken us into the Anthropocene 222
We have the opportunity to live better than ever 222
The low carbon technologies we need are coming along nicely but on their own they won't help 222
Anthropocene challenges are global, systemic and inescapably intertwined 223
We need to stand further back from the problem and this entails slowing down more of the time 223
We need a new system of economics fit for the twenty-first century 223
Some types of growth are still healthy but others are not 224

We will require globally shared values of respect for all people, for the planet, and for truth 224
We humans urgently need to develop our thinking skills and habits in at least eight respects 224

12 WHAT CAN I DO? 225
How can I help to create the conditions under which the world that I want to see becomes possible? 225
What questions were missing? What answers were wrong? 227

Appendix: Climate Emergency Basics 229
Alphabetical Quick Tour 242
Notes on Units 270
Endnotes 273
Index 305

ACKNOWLEDGEMENTS

A small army of people helped to dig out data, create graphs, edit text and provide ideas. Many thanks to Sam Allen, Kez Baskerville-Muscutt, Kirsty Blair, Sarah Donaldson, Kate Fearnyough, Lucas Gent, Cara Kennelly, Moe Kuchemann, Cordelia Lang and Rosie Watson.

Thanks also for advice, comment and encouragement from Myles Allen, Nico Aspinall, Libby Davy, Pooran Desai, Robin Frost, Magnus George, Chris Goodall, Chris Halliwell, Phil Latham, Tom Mayo, Rupert Read, Jonathan Rowson, Stewart Wallis, Nina Whitby and Gary White. Nick Hewitt and I worked together on the food papers that are central to the first section, with help from some of the army listed above.

Lancaster University has been a fabulous boiling pot of interdisciplinary ideas. Thanks especially to everyone in the Institute for Social Futures and all those who chipped in their thoughts at the Global Futures events, where nearly all the themes in this book have had an airing.

Thanks to Matt Lloyd and all at CUP for confidence in the concept, patience, and of course great edits.

For this revised edition, I am also grateful to the many hundreds of people who emailed me with comments and feedback on the original edition.

All the mistakes are mine.

Three sheds let me escape from the noise and haste, and were critical for perspective. One is at the end of my garden. One is an off-grid metal box on Colonsay, my favourite Scottish island. The third is the unforgettable sunrise on the Cumbrian West Coast, to which Chris and Elaine Lane very kindly lent me the key.

Most of all, as always, thanks to Liz, Bill and Rosie for putting up with the whole process and dad's grumpy pre-occupation.

WHAT'S NEW IN THIS UPDATED EDITION?

I'm writing this in lockdown, as coronavirus disease (COVID-19) holds humanity in its grip. Everyone is trying to work out what it will mean for the world after the immediate crisis has passed. Many people are asking whether this could be the moment we unlock ourselves from an economic and social set-up that is now so clearly unfit for the modern era and embark, at last, on the transition to a more environmentally friendly world that our (and other) species so urgently needs. Of course, at present no one knows. But clearly, change of every kind, for better and worse, looks so much more possible now that we have already been forced to turn our lives upside down.

Even without the pandemic, the two years since publication of the first edition of *There Is No Planet B* have seen a lot of change. In 2019–20, the wildfires of Australia single-handedly added more than 1% to global greenhouse gas emissions and turned the skies orange in New Zealand, 2,000 miles away. Methane has been exploding (literally) out of the melting Russian permafrost, leaving craters 50 metres across, and we have news that the Amazon may be drying out so fast that it could soon turn from an important carbon sink into a brand new source of emissions.

And yet at the same time we have also seen hugely encouraging signs that the world – or at least some parts of it – is at long last waking up to our environmental crisis. There is such a long way to go, but it is enough to give me more hope than I've had in years.

When the first copies of *There Is No Planet B* reached the bookshops, the Intergovernmental Panel on Climate Change (IPCC) had only just released its long-awaited report, making crystal clear the need to keep global temperature change to within 1.5 °C.[1] Greta Thunberg was only just emerging as a household name, and the school kids had not yet taken to the streets *en masse*. Extinction Rebellion was still planning the landmark London protests of April 2019. In the UK, politicians still thought they could look like they were leading the world with a hopelessly inadequate target to cut emissions by just 80% by 2050. It had not yet become fashionable for both local and national governments to declare a climate emergency. The BBC was only just starting to talk plainly about the need to eat less meat and dairy. It had not yet released its landmark documentaries *Climate Change – The Facts* (which drew a line under its hitherto abysmal coverage of the climate emergency) and *Extinction: The Facts*, which brought our biodiversity crisis into shockingly sharp focus. (Please try to watch these documentaries if you haven't already.)

Plans to expand Heathrow Airport have been judged illegal by the Court of Appeal on climate grounds. That sets an incredibly helpful precedent, opening the floodgates for climate-scrutiny of all the UK's infrastructure plans, including crazy new coal mines and road expansion schemes. It is getting harder for politicians to talk about the climate in one moment and then forget about it the next. These straws in the wind are enough to give me real hope.

How did these first steps of change become possible? What caused what? The answer has to be that, just like all system change, it all came together at once. The *conditions became right* for the first inch of shift to take place on the big transition; the journey that we so urgently need. And the first inch might have been the hardest one.

It is a race between tipping points; will the environmental crisis tip out of control first, or will humanity wake up in time? While the science is getting even more scary, for the first time, it feels like society as a whole has actually managed to stir a little in its slumber. Our collective head has just started to try wiggling out of the sand. If we all pull hard, the big system change that we so urgently need feels more likely than it did two years ago. And we might even be in time.

COVID-19 struck at a moment when the world was tottering on the brink of enormous change. By delivering an immediate threat to every human life, the pandemic has surely forced us to reflect on what truly matters in life. Lockdown gave humanity a chance to stand back, take stock and rethink. This book tries to help with that by looking at how the old system was working, what needed to change, why and how.

So, what's specifically new in this edition? I haven't found any need to update the basic premise or through-line of the book, or even any of the numerical analysis, but the launch of this new version gives me a chance to fill in some gaps and cover some critical updates.

To start with, of course, the whole book has been picked through, nuances and updates squeezed in all over the place and some of the diagrams improved. In quite a few places I have replaced the phrase 'climate change', with a phrase that better describes what we now face: 'climate emergency'. In the light of COVID-19 I have written more about disease. I have considered around 700 feedback emails, making useful tweaks and insertions throughout. So this version of *There Is No Planet B* is even more of a collaborative work than the first version.

In terms of brand new material, I have written more about some of the important levers of change that have been emerging. There is now a short chapter on protest, including Extinction Rebellion and the striking school kids. I've also added more for the business community in response to the surge of new questions that I am getting in my work: Does offsetting work? What does 'net zero' mean? How should we invest? How should we manage our financial asset portfolios to get the trillions of dollars pushing for a better world? Finally, in recognition of an even stronger calling for all of us to be part of the change right now, I have elaborated on the 'What can I do?' section.

We are all on a journey of change now and I hope that this new edition reflects that more than ever.

Happy reading!

Mike Berners-Lee

INTRODUCTION TO THE FIRST EDITION

Welcome to a new era

Almost every year since records began, our species has had more energy at its disposal than it had the year before. For the past 50 years, the growth rate has averaged 2.4% per year, more than tripling in total over that time. For the century before that it was more like 1% per year, and as we go back through history, the growth rate looks lower still but nevertheless positive, give or take the odd blip. We have been getting continually more powerful, not just by growing our energy supply, but by using it with ever more efficiency and inventiveness. In doing so, we have been increasingly affecting our world, through a mixture of accident and design. The restorative powers of our planet, meanwhile, have remained broadly the same, so the *balance* of power has been shifting – and it has now tipped. Throughout history, the dominant cultures have treated the planet as a big and robust place, compared to everything we could throw at it – and that approach has not, generally speaking, come back to bite us.

But sometime in the past few decades, things changed. We can argue about exactly when, but let's just say that it happened recently. Around 100 years ago, in the First World War, we couldn't have smashed the whole place up even if we'd tried. But 50 years ago, with nuclear energy especially, it became clear that we could totally mess things up if we made big enough blunders. Today, we don't have to blunder at all; if we don't try hard enough NOT to we will wreck the whole environment. And 50 years into the future, if the energy-use trends continue, the world will be more fragile still, compared to our ever-increasing might.

For another perspective on human energy growth, think back to 26th December 2004, when Asia experienced a tsunami that killed 230,000 people. It was a big natural disaster that most people reading this book will remember. The energy released in that wave was about equivalent to 24 hours' worth of human global energy consumption at the time. One hundred and fifty years ago, it would have taken humanity about a month to acquire and use the same amount of energy. Today it only takes us 18 hours.[1]

The 'big people, small planet' syndrome has a name that is going to be useful to us: the **Anthropocene**. I use this simply to mean the era in which human influence is the dominant source of change to the ecosystem.

Our arrival in this 'Anthropocene' has been like a pH titration experiment. In the lab, acid might be dripped into a flask of alkali solution. For ages there is no colour change at all because the alkali still dominates, then suddenly, one more drip and the balance shifts. The flask turns acidic, the indicator turns from blue to red and the world inside the flask becomes an entirely different place. In our global experiment, we have been adding more and more human power into the mix but for millennia the planet's restorative power still dominated. Although we wiped out some other species, we have broadly got away with treating the world as a big sturdy playground. Suddenly it is fragile. The playground will break unless we dramatically change the way we play in it. And this particular titration experiment has also been a crazy one because whereas in the science lab the closer you think you are to the balance point the more slowly you add the acid, we are pouring our power on faster and faster.

In the past, humans have always been able to expand as they have developed, but suddenly now, and for the foreseeable future at least, we can't. That is a massive change. Even for those who are starting to view the one planet constraint as temporary (and I'll debunk this later), the physicist Stephen

Hawking put it like this: 'We will not establish self-sustaining colonies in space for at least 100 years, so we have to be very careful in the meantime.'[2]

There is no Planet B.[3]

A handbook of everything

This is a book about the big picture of life on our small planet. It is an evidence-based practical guide to the make or break choices we face now. It is about taking the chance for us to live better than ever and heading off the threat of living worse or not at all. It is about the difference any of us can make, despite the global nature of the challenges.

A few years ago, all my work focussed on the climate emergency. That wasn't because only climate change mattered, but because at the time it seemed like a useful and practical simplification to treat this one component of the Anthropocene challenge in relative isolation from the rest of the rich soup of environmental, political, economic, technological, scientific and social issues of the day. But it became increasingly and inescapably clear to me that the climate emergency had to be dealt with as a multi-disciplinary challenge.

It also became ever clearer that while climate breakdown presents a tangible environmental problem, it is not the only one, and won't be the last. We have had decades of warning about climate change. But we have wasted that time through our denial, first of the problem itself and then of the nature of the solution that is required, and through the unspeakably clumsy way in which we inch towards the kind of global agreement that might actually help. In the Anthropocene, we can't rely on every challenge giving us so much warning. We'd better practise our global governance because we might need to respond to something just as intangible as climate change on a far shorter timescale. What exactly? That's the point; we don't know yet. One of the key things we need to understand is just how much we don't know.

I have called this a handbook because it is intended to inform decision making at every level, from the individual to governments. Everyday tips are mixed in among messages for policy makers, voters and business leaders. Some relate to what I would call the 'intensive care' end of things; how we can manage the known and tangible challenges such as the climate emergency, food security and biodiversity. Mixed in with this, and inseparable from it, are the deeper underlying questions of how we can get better at heading off these kinds of challenges in advance; questions of 'long-term global health'.

I hope you enjoy the perspective-forcing facts, stats and analysis, some of which surely have to make us all gasp at the way we live, and the opportunities to do better.

I have taken on just about everything at once, simply because no other approach will do. It doesn't work anymore to look at technical questions of food, energy or the climate emergency separately from each other or separately from questions of values, economics or the very ways in which we think. All these things are too inescapably intertwined for the traditional 'one bit at a time' approach to be adequate. We must look at all these complex problems concurrently, and using a range of disciplines or *lenses*.

For this book, that means we are going to move from the very big picture to the specifics and back again, and from one discipline to the next as the need arises. I hope this makes for an interesting ride.

In over a hundred talks, workshops and seminars, surely by now I've been asked every question it is possible to think of: 'Who should be leading on this?'; 'Are humans too fundamentally selfish to deal with the climate emergency?'; 'If I don't fly, won't someone else take my seat?'; 'What's the point in me doing my bit when I'm just one speck among seven billion?'; 'Do we have to halt economic growth?'; 'Does it all boil down to population?'; 'Why should I bother when we all know we've had it anyway?'; and so on. I've felt naïve jubilation at the simple freedoms and opportunities that a low carbon world could have. And in writing *The Burning Question* with Duncan Clark, I have fought through the gloom that came from contemplating deeply, day after day, how far away almost everyone seemed to be from grasping what we came to see as the essential basics of the climate emergency, let alone confronting the issue properly.

And since then I've seen small but real steps that provide a lot more grounds for hope. I've reflected and jostled with the dilemmas and hypocrisies of my own lifestyle. I've felt foolish at the futility of not flying and guilt at doing so. I've loved the cycling and the excuse to be scruffy at work and I've also been sobered by my friend's head injuries after he tumbled off his bike on his commute home. By now, surely, all the world's biggest dilemmas and conflicts have already gone on in my own mind? Of course, I know that's not true. But let's just say that I've done quite a lot of thinking, talking and sense-making. I have picked the brains of a lot of smart people. And now, with the help of as much collaboration as possible, it is time to put it in a book.

This book is about how we can make the transition into a new mode of living that works for us and fits our new context – a way of operating that won't smash the place up and will allow us to thrive despite our power.

When it's all so global, what can *I* do?

This is one of the crunch questions of our time.

While our collective power has been rising, so too has the population, and each one of us is becoming an ever-smaller part of the whole. It is easy to feel like a speck or an ant in the unstoppable trajectory of humankind's time on the planet Earth. It can be tempting to think that, whether or not we like where we are heading, there is little influence that any of us can have.

It is a valid concern. As we will see later, at the global system level, there are powerful feedback mechanisms at work that have so far been immune to most of the efforts made not just by individuals but by organisations and even nation states. At the moment, humanity can be seen as slave to a dynamic interplay of growing energy, efficiency and technology, a persistent trajectory whose broad trend we have so far shown little or perhaps no ability to influence. To give just one example, it is a stark truth that the sum of all the world's climate action has so far made little or perhaps even zero detectable impact on rising global emissions. Ever more energy and technology may have brought us many good things so far, but quite suddenly it has become dangerous to continue on the same pathway without greater control. And to achieve that, our species will need to both raise and *change* its game.

We urgently need new problem-solving *methods* from those that were honed in the days when the nature of the problems was altogether simpler. But changing the way we think isn't simple because we are dealing with grooves of habit that have been worn deep over the centuries.

One way of looking at it is to say that we need to *rebalance* our evolution. Our technological brilliance has taken us into a situation in which we need to quickly evolve in other ways as well. Life can be better than ever before but that won't happen unless we can balance our technical genius with the development of some very different and complementary thinking skills to go with it.

Is our failure so far to take the reins proof that we are powerless to do anything to affect things at the global system level? I don't think so, but in this book, I am going to take that question seriously. We are going to explore the big system dynamics and ask what they tell us about how the individual can actually be useful. I think we can each have far more impact than most people assume, but we need to get a lot smarter at understanding which kinds of things make a difference and which don't. We need to think beyond the immediate and direct effect of our actions and ask more about the ripples that they send out, and how the actions of one person, company or country might get multiplied rather than muffled or counterbalanced by the rest of the system.

Although it runs against some of my instincts to tell anyone what to do, this book contains plenty of suggestions. I have done this because it is so easy to think there is nothing any of us can do and I want everyone to see how that is not the case. Often my suggestions are very simple. And don't worry, this isn't a lifestyle guide for perfect people; I'm far from one of those myself and don't expect you to be one either. But like me, you care a bit and you are interested to know more about what makes sense on every scale, from the personal to the global. So, I hope you find some stuff here that you can use.

What values underpin this book?

This critical topic of values gets a section of its own towards the end, where we will look from a purely practical point of view at which values can and can't help us to thrive over the next hundred years. But for now, I will just note a few that underpin the book. If you can't live with these, or something fairly similar, then there might be little point in you reading on – so at least I've saved you some time.

I have written from the perspective that all people have equal intrinsic value as human beings. Rich, poor, black, white, American, European, African, Chinese, Syrian, Muslim, Buddhist, Christian, Atheist – *everyone* has the same intrinsic value. For many, this might sound too obvious to be worth writing down, but all too often the values behind ideas are not made explicit even though the implications are usually enormous for economics, food policy, climate policy and just about everything else you can think of for thriving in the Anthropocene. To be very clear, the same principle of inherent equal value of all human beings is universal. It applies to all world leaders, purveyors of both real and fake news, tireless aid workers, left wingers, right wingers, billionaires, paupers, your own kids, other people's kids and even the call centre employee who rings you when you are having dinner with your family to try to persuade you to sue for an accident that never even happened. A person's inherent worth is independent of their circumstances or the choices they have made in their lives or have had made for them.

As far as other life forms go, they too deserve a place – because of their own sentient experience, not just for the practical reason that humans need them for food and medicine. My son asked me about the relative value of a woodlouse and a human embryo of the same size. I can't answer that question. Let's just say, for the purposes of this book, that all forms of life matter.

I'm going to try to stay consistent with these values throughout, but it won't be easy, since so much of what goes on, in both everyday life and in politics, is actually in stark contradiction with this simple principle. And just to be clear, you don't have to dig all that hard to find that plenty of my own life conflicts with these values too – I'm no more of a saint than anyone else is.

If you do broadly agree with these values in principle, then let's explore a few implications. It means that while you might want good things for your own country, you don't want that at the expense of other countries. If you want your country to be 'great' or even 'great again', you would be careful not to go about engineering that at the expense of any other country's 'greatness'. It means that if you want the best for your kids, the way you go about it is not at the expense of other people's kids. It means that if you are in hospital trying to ensure that your elderly parents don't die unnecessarily, you are not trying to ensure they get the best care if it is disproportionately at the expense of resources for others who need it just as much (and this is challenging – I've been there). It means, for example, that if you were given the chance to vote on whether or not your country should be in the EU, you would be thinking not only about your own interests but about the interests of your whole country as well as the interests of the whole EU *and* the wider world. It means that, when you go shopping, what you buy is not just the product itself but also a whole set of implications for everyone who was involved in producing it. This is the hidden stuff, almost entirely ignored by the advertising industry, which we need to find a way of tuning in to.

What can we aim for?

Is it possible to have a universal vision that floats nearly everyone's boat? While the idea of limiting climate breakdown seems like essential damage limitation, in itself, it spectacularly fails to excite most of us. More often than not, it gets framed primarily as the need to forego things we enjoy. And since humans – all of us – hate thinking about anything unpleasant, the temptation to switch off is hard to resist. Like it or not, that's how our psychology works. Unbelievable fantasies about the future don't work either – creating a sense that what is worth having is impossible.

Luckily, there is plenty of scope for realistic improvements that are well worth getting excited about. So far, we have not exactly managed to optimise the quality of human experience. Dealing with the big issues gives us a chance to reengineer things for the better. We don't spend enough time imagining good futures, so we end up on a 'business as usual' pathway, just because we haven't really thought properly about anything better.

I don't want to prescribe too tightly because we all see things differently (thank goodness), but here is my attempt at sketching out what I think we can aim for and most of us might want. I'm not anticipating perfection, but the closer we get, the better life will be, and even trying to head in this direction should be a good experience.

Here goes. This book is geared towards a future along these lines.

The air is fresher. Life is healthier, longer, more relaxed, more fun and more exciting. Our diets are varied, tasty

and healthy. More of us get out as much as we want to, both socially and physically. Travel is easier – but we spend less time in transit. We feel freer to live life in whatever way seems meaningful to each of us at the time, in negotiation with other people's equal right to do likewise. There is less violence at every level. Cities are vibrant while the countryside teems with wildlife. Our jobs are more interesting, and the pressures are more often self-imposed. We expect, insist on and get higher standards of trust and truth, in politics, in the media and in fact everywhere. We are better connected to the people around us and to our sense of the global community. We give more of our time and attention to others and we notice and enjoy more of what is going on around us. We might compete with each other for fun but where it really matters we collaborate better than ever before.

Of course, within this, there is plenty of fleshing out yet to be done, and enormous variety in exactly how each of us will live within it. Please feel very free to add details and perhaps think about where you personally might want to fit into life on Planet A.

Not the last word ...

A lot of this book is just about laying out evidence that largely speaks for itself, but where I've made interpretations I hope you find they make sense. Of course, I don't expect to have written the final word on any subject; all I offer are rough outlines, waiting for improvement. I hope they are a sensible start point. If I've got something wrong, I hope you will notice. I hope debates will rage, as they urgently need to. I will be delighted by anyone who rejects anything I've written in favour of something better, and where you think I've missed bits out, I hope you'll fill in the gaps, so that by the end, your understanding will be better than mine. Please send constructive feedback and improvements to Mike@TheresNoPlanetB.net.

However flawed my attempt, I am confident that it is better to have a rough sketch of everything at once than to see things through only one lens at a time, however perfect each view.[4]

If you use Twitter, #NoPlanetB would be one way to share ideas.

1 FOOD

How humanity is fed now and how we can do this better in the future. What can be done and what can everyone do?

We will start our tour of the big picture by looking at the global food system since food is the original source of energy for humans and is still as essential as ever.

Our land and sea need managing from many different perspectives at once. We need to feed a growing population with a healthy, tasty, low carbon diet. But we need to achieve this while preserving or improving the biodiversity that is currently haemorrhaging and despite the reductions in land fertility that we may be causing, not least through climate change. We also need to fend off pandemics, a looming antibiotics crisis and an explosion of plastic pollution that has crept up on us in just the past 50 years and is now with us forever as far as we can tell. As if all this wasn't enough, even though we don't really know how to do it yet, it is becoming increasingly clear that we will need land to have a role in putting carbon back in the ground. Oh yes, and we also need it for living space and recreation.

And these are just the human-centric considerations. How many readers will write me off as a hippy if I mention again that animals might also matter as sentient beings?

Luckily, for all the hideous complexity, it turns out that some relatively simple analysis makes a few important things very clear. Whether you are a food policy maker, a producer, a retailer or just someone who eats food, here are some big messages that I think everyone needs to know. They tell us a lot about what we can all do to help.

How much food energy do we need to eat?

About 5% of all human energy is still consumed in the most traditional way of all: through our mouths. On average we need 2,350 kcal (calories) per day, but we actually eat about 180 kcal more than that.[1]

The average requirement of 2,350 kcal per day takes account of the different ages, genders, sizes and lifestyles of the world's population. It works out at 114 watts. For comparison, a big plasma TV needs a similar amount of energy and an electric kettle gets through about 15 times as much when it's on.[2]

How much food do we grow worldwide?

At the global level, we grow 5,940 kcal per person per day. That's nearly two and a half times as much as the 2,350 kcal per day that the average person needs to eat to be healthy.

Given these stats, you would think the planet Earth should be the land of food-plenty.

Regional variations are huge. North America grows a massive eight times its calorific requirement. In Europe and Latin America, the food grown is 'just' four times what humans need to eat. But sub-Saharan Africa grows only one and a half times the calories it needs.

You may well ask: why on Earth does anyone go hungry, and what do the Americans do with all their calories?

To find the answers to these questions we need to have a proper look at the journey from field to fork.

What happens to the food we grow?

Some 1,320 kcal per person are lost or wasted, 810 kcal go to biofuels and a massive 1,740 kcal are fed to animals.

(But meat eaters relax and read on. You don't need to go 100% vegetarian or vegan unless you want to.)

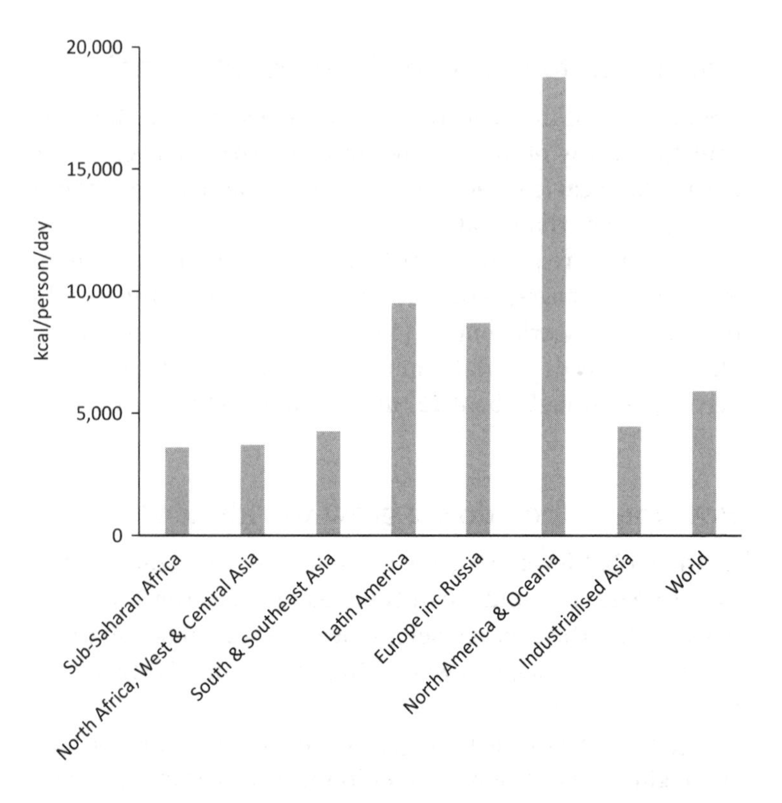

Figure 1.1 Plant-based human-edible food grown per person per day in different regions worldwide.

This simple chart cuts through the complexity of the global food and land system to give us a startling and essential perspective.[3] All the numbers are in calories per person per day. I thought I already knew quite a lot about sustainable food having taken a keen interest for a decade or so, but crunching these numbers properly for the first time recently was a revelation.

Of the 5,940 kcal grown per person every day, here is what happens. Right at the start of the journey from the field to the stomach there are two kinds of waste; 340 kcal don't even get harvested. Some of that is down to over-fussy quality standards in developed countries, or gluts where supply exceeds commercial demand and the food is shockingly left in the ground. Most, however, is down to harvesting inefficiencies, and while there is room for improvement, harvest loss is impossible to eradicate altogether. A further 330 kcal or so get lost in storage. This is mainly a problem in poorer countries and is often simply down to the lack of a sealed, dry container. At face value, there is plenty of scope for cutting this down.

Even after these stages, there are still a massive 5,270 kcal which are allocated in four key ways.

A small amount, 130 kcal, is replanted. This is a good idea because it means we can eat next year as well. And 810 kcal go

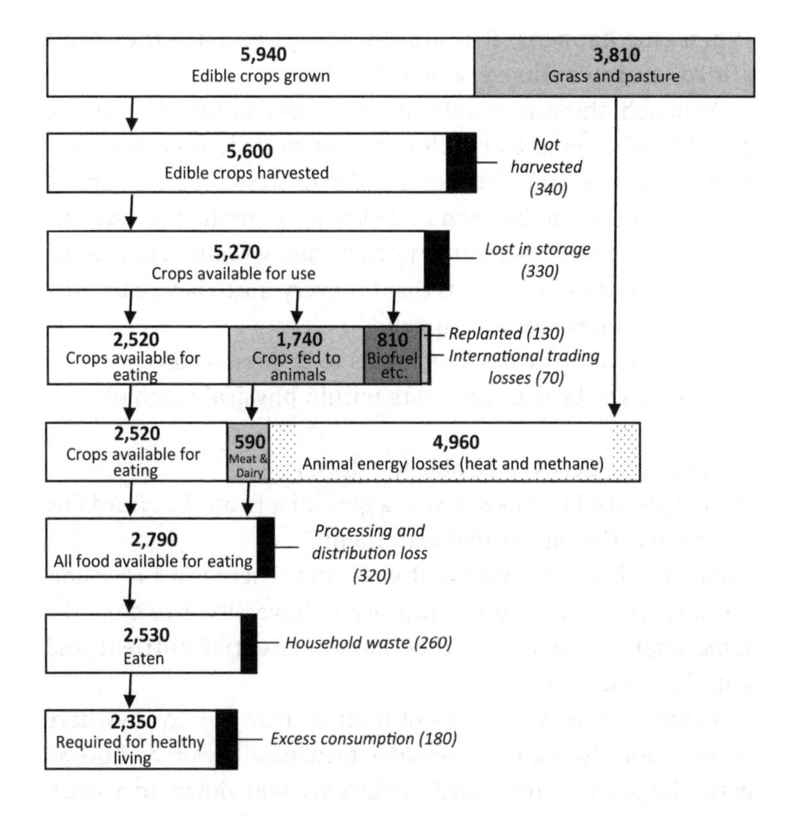

Figure 1.2 The world's food on its journey from field to stomach. Numbers are in kcal/person/day.[4]

to non-food uses, which is mainly biofuel. Animals eat a massive 1,740 kcal. That still leaves 2,520 kcal of plant matter for human eating.

After that there are some relatively small losses in distribution and food processing, and then households waste a good chunk more. In the end, including the meat and dairy, the average human eats 2,530 kcal, which is 180 kcal more than the average person needs for a healthy diet.

Given the global surplus, why are some people malnourished?

When this happens, it is almost always because they can't afford, or don't choose, a healthy diet.

Although there is significant net over-consumption at the global level, around 800 million people go undernourished (not enough calories) and a further 2 billion or so suffer some kind of 'hidden hunger' in the form of deficiency in protein or essential micro-nutrients, most commonly iron, zinc, vitamin A or iodine.[5]

One way of looking at it is that for everyone to have a healthy diet, four things need to happen.

(1) Enough of every nutrient needs to be produced.
(2) It needs to be transported to within physical reach of everyone.
(3) Everyone needs to be able to afford it.
(4) People need to choose to eat a good diet from the affordable options that are available to them.

Today, the first condition is already met. There is a 14% calorific surplus. With others at Lancaster University, I've done the same analysis for every other human-essential nutrient and found a similar story.[6]

Supply chains are capable of finding their way to anywhere in the world that can pay to make it financially worthwhile. So both the second and third conditions boil down to wealth distribution, which we will look at in more detail later on (see pages 130–139).

Taking all four criteria together and given today's food supply and population, there are just two critical factors for a healthy diet for all: it boils down to money and choice. Inequality is the main reason why anyone today does not have access to a healthy diet. Without sorting this out, it looks highly likely that there will always be hunger, whatever happens to the total supply. To be clear, when we come to look at wealth distribution we will find that the problem is to do with *relative*, not *absolute*, wealth.

Choice is of course a complex issue, combining education, culture, mental health and personal taste.

Importantly, nobody starves today for lack of food at the global level. The current problems are about how the abundance of nutrition is shared around.

Why don't more people explode from overeating?

Luckily, overweight bodies are less energetically efficient. Otherwise, many of us would.

If all of the net over-consumption of 180 kcal per day were translated into extra body weight, the average person would be gaining weight at about 8 kg per year.[7] This would be disastrous in just a few years. Luckily, as a body gets overweight it gets less efficient, and burns through more energy just getting through the day.

However, if all humans were a healthy weight and eating only what they needed to maintain that, it would liberate food for a billion or so of the extra mouths that are coming our way.[8] Clearly there would be other wellbeing benefits in parallel. Easier said than done, I know.[9]

We turn now to a closer look at the role of animals, which leaps out from the global food calorie flow map as a cause for concern.

How many calories do we get from animals?

Animals contribute 590 kcal to the human food chain as meat and dairy. BUT they eat 1,740 kcal per person per day of human-edible food as well as 3,810 kcal of grass and pasture.

The average farm animal converts just 10% of the calories it eats into meat and dairy foods. The rest is used up doing things like keeping warm, walking around, burping up methane and creating dung. While more than two thirds of all farm animal food are grass and pasture, which cannot be eaten directly by us, the human-edible crops that we feed them amount to more than three quarters of the calorific needs of the entire human population.

We can't eat grass and pasture, but some of the land currently dedicated to its production could be used for crops and some of the rest could be very usefully set aside for biodiversity.

In terms of efficiency, two rules apply. Firstly, the conversion rate is better when you don't kill the animal but instead take its eggs or milk. Secondly, if the animal doesn't have to keep warm, move around or live long then less energy will be wasted. Hence the efficiency of conversion is particularly low for beef (typically ~3%) but highest for eggs and milk (~18%). Obviously, and inadequately, this analysis has ignored any consideration of an animal as a sentient being.

How much do animals help with our protein supply?

They don't. The world's farmed animals destroy nearly three quarters of the protein that they eat, most of which comes in the form of human-edible food.

The average human needs about 50 grams of protein per day for a healthy diet and this is one argument that is sometimes used to defend the world's growing meat and dairy industry.

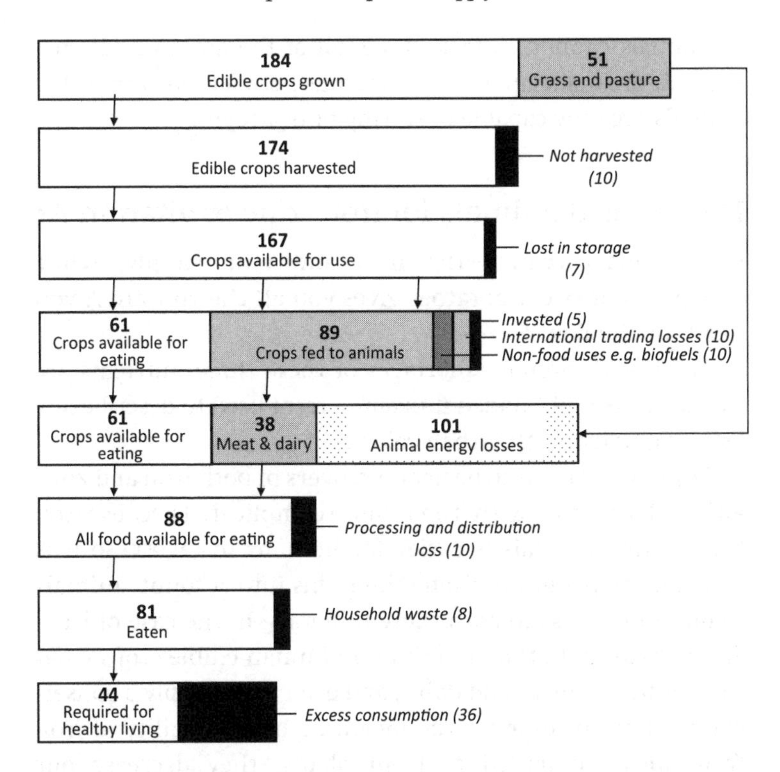

Figure 1.3 The journey of the world's protein from field to stomach.[10] Numbers are in grams of protein per person per day.

Protein can be tracked from field to fork in the same way as calories. When we do this, we find that a few myths can be busted. Firstly, we would have far more protein if we didn't feed human-edible plant protein to animals. Secondly, the world has an even greater surplus of protein than it has calories. This last point is made more complicated by the fact that it is somewhat harder to distribute protein evenly than it is calories. Calories, unlike protein, are self-regulating to some degree. If a person regularly eats twice the calories they need they become extremely unhealthy in a very short time, whereas you can do the same with protein and not even know you are doing it.

Animals cannot, in fact, create all of the amino acids that proteins are made up of. There are nine *essential amino acids* that animals are only capable of storing or destroying.

Do we need animals for iron, zinc or vitamin A?

No. Animals reduce the iron and zinc supply, while 100 grams of sweet potatoes gives you all the vitamin A you need for the day.[11]

Along with protein, shortages of these three nutrients are the key causes of 'hidden hunger', a term used to describe any lack of nutrients other than calories.[12]

Animals turn out to be net destroyers of both iron and zinc, although the story with iron is more complicated because iron from eating animals is easier for humans to ingest, so it is worth more per gram. Even taking this into account, animals reduce our access to both these minerals. In the case of iron, animals eat more than ten times in human-edible crops what they return in meat and dairy, and even if you apply a conservative factor of four for the increased bioavailability of iron from meat compared to from plants, they decrease our supply. Meanwhile, less than one fifth of the zinc that animals take from human-edible crops comes back to us in meat and dairy.

Vitamin A is a bit different. It is one of the very few human-essential nutrients that animals create more of than they consume from human-edible crops. So in the days before it could be manufactured and routinely added to foods, the need for it could have provided a legitimate argument for more poultry and dairy in the diet. However, the advent of fortification has changed things a great deal. Vitamin A is easy and cheap to add to oils and even flour, and fortification is routine in many countries, including the UK and USA. Interestingly, China would be one of the few countries to have a net surplus even without fortification or meat and dairy, thanks to its sweet potato supply. If you are concerned about vitamin A in your own diet, just 100 grams of this delicious vegetable gives you an amazing 709 micrograms of the stuff (compared to a recommended daily intake of 700–900 micrograms). And because they travel well on boats, sweet potatoes make a great part of a sustainable diet wherever you live. Although not as spectacular, carrots, olives and most leafy green vegetables are also very good.[13] Finally, if none of these solutions do it for you, popping a pill is a simple and cheap last resort.

As a general rule, animal products are not a twenty-first century solution to micro-nutrient concerns, except perhaps in parts of the world that both lie outside the global food economy and are without access to proper health care. For these areas, it is still true to say that in the absence of access to a varied diet or supplements and fortification, a bit of meat can be a simple way of catching up on a variety of essential micro-nutrients. However, those circumstances will not on the whole apply to those who have access to this book.

How much of our antibiotics are given to animals?

An estimated two thirds of all antibiotics,[14] 63,151 tonnes per year in fact,[15] are gobbled up by animals – and some of that even makes it back to us through meat and milk.

One of the most powerful arguments for the benefits of modern technology is the increase in health and life expectancy that most of us can enjoy. The collapse of antibiotics would send a lot of that down the plug hole. And it looks close. The race between increasing resistance and the development of next-generation alternatives looks like it is going the wrong way with extremely nasty and perhaps imminent consequences. (This feels especially real for me because without them over the past five years, I might well have died, in a very unpleasant way too, both my parents would have died for sure, and there is a very good chance my daughter would have had her leg chopped off or worse.) If you find yourself in serious need of an antibiotic, the idea of a world without them turns from a distant concept into a vivid nightmare.

Animals are given antibiotics mainly to stimulate growth and prevent rather than cure disease. This quantity used worldwide is going up fast, as diets in developing countries are changing in the wrong direction and farming practices are intensifying. The result is that animals are developing resistant strains and passing those bugs onto us. We can't blame it all on the farmers though, because much of human consumption of antibiotics is also needless.

What can I do and what can be done?

The World Health Organization offers basic advice for all.[16] Here are some key points, plus my own comment on diet.

- Don't take antibiotics unless you need them, and when you do, follow the instructions.
- Cut down on meat and dairy that comes from farms that routinely use antibiotics for prevention rather than cure (and it is fairly safe to assume the worst unless you know otherwise). The 'organic' criteria include restrictions on antibiotic use.[17]
- Have good hygiene and keep vaccinations up to date to prevent infection in the first place.
- Farmers should not use them for growth stimulation or disease prevention, and should use vaccinations and good farm hygiene to prevent disease.

Do factory farms make pandemics more likely?

Yes. It looks very likely that today's farming practices facilitated COVID-19.

I write this as coronavirus disease (COVID-19), the latest disease to jump from animals to humans, still holds the world in fear. There have been several pandemics over the centuries caused by diseases jumping to humans from other species, most notably the Ebola virus and the human immunodeficiency virus (HIV), as well as 15 influenzas over 500 years which came from birds. And the frequency has been going up. COVID-19 shows some but not all of the qualities of a disease capable of inflicting a full-scale disaster so it looks as if on this occasion only a small proportion of humanity will have been wiped out. It seems to be very contagious, even before the symptoms emerge, but so far, at least, it is far less vicious than, for example, the Ebola virus or the SARS virus of 2004. So you could say we have got off lightly this time and should think of it as a wake-up call.

As I write this, many people, including me, are trying to work out whether this could be the moment that humanity starts the major re-wiring project that we so badly need, but for the moment, within this food chapter, let's just ask: What has our diet and our farming system got to do with it all?

We don't yet know all the details of how COVID-19 came about, but there are some clear factors that increase our risk from pandemics.

First of all, animals crammed together in huge numbers have a capacity to rapidly cause a virus to mutate into a more dangerous form. This is a feature of factory farming, especially but not exclusively for chickens, where the numbers of animals crammed together are so high.[18] The close cloning of a species that takes place in industrial farming in order to improve the quality and yield of the meat further facilitates the spread of viruses between animals and the ease of virus mutation. A further outcome of efficient factory farming has been the emergence of wild-animal farming in China, as small farmers have been pushed out of conventional meat markets. In the case of COVID-19, as I write, it looks likely that the origin may have been bats and the jump to humans came via the intermediate stage of another farmed mammal.[19]

So what is the solution? Less meat in our diet, more care over how it is reared. More space between animals. Less tight cloning between species. Better regulated meat markets around the world. Less trade in endangered species for 'homeopathic' medicine.

How much deforestation do soya beans cause?

Don't blame the soya bean! The problem comes when they are eaten by cows and sheep.

Gram for gram, a soya bean has more of almost every human essential nutrient than beef or lamb. But when you feed one to a cow or a sheep, you only get about one tenth of the weight back in meat. It's a disaster for human nutrition. The bad reputation that soya beans get for causing trees to get chopped down is misplaced.

The second myth about soya beans is that they don't taste good. They do, either as milk, tofu or simply as the beans.

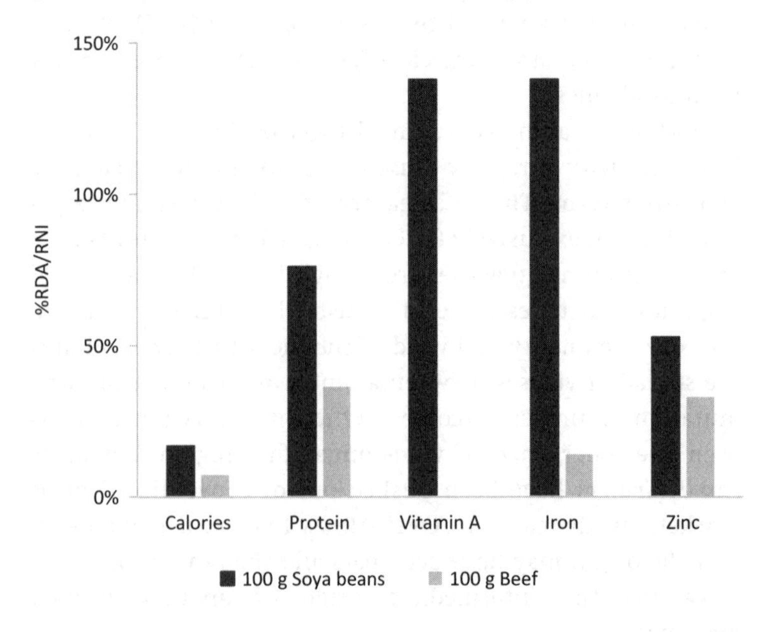

Figure 1.4 The calorific, protein and micro-nutrient content of 100 grams of soya beans and 100 grams of beef, in terms of percentage of Recommended Dietary Allowance (RDA) and Reference Nutrient Intakes (RNI).

What's the carbon footprint of agriculture?

At 23% of the global total, food and land-related emissions are far too important to ignore.[20]

Most people who care deeply about the climate emergency seem to get so exhausted trying to make sure we leave the fossil fuel in the ground that they don't have much energy left to look at food and land. It is understandable, but unsatisfactory, since food and land emissions are enough on their own to see us in climate trouble. They are the forgotten poor relation in the climate change debate.

In rough numbers, humankind's greenhouse gas footprint is 50 billion tonnes of carbon dioxide equivalent (CO_2e) per year, of which about 23% comes from food and land. Agriculture's single biggest source of CO_2 is deforestation, most of which can

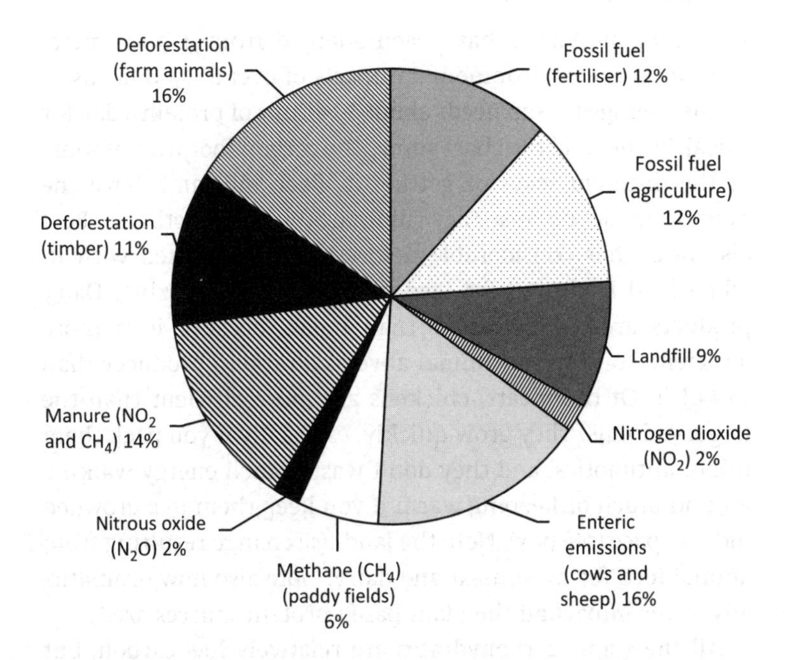

Figure 1.5 Breakdown of food and agriculture's 23% contribution to humankind's greenhouse gas (GHG) footprint.

be tracked to meat production, but some to timber. When the trees are cut down, we lose not only the carbon stored in the wood, but even more importantly, over a period of a few years we lose most of the carbon stored in the soil as well. Fossil fuels used in the manufacture of fertiliser, powering agricultural equipment and in transport, are a relatively minor player. But on top of the CO_2, most of the world's methane can be put down to food and land, with the greatest sources being enteric emissions (that's cows, sheep and goats chewing the cud and burping), flooded paddy fields, and rotting food matter in poorly managed landfill sites. About two thirds of all nitrous oxide is also attributable to food.

What are the carbon footprints of different foods?

The following charts have been adapted from a huge meta-analysis of the environmental impacts of over 38,000 farms.[21]

The average person needs about 50 grams of protein a day for a healthy diet and the chart shows the carbon footprint of some of the different ways of getting it. Beef and lamb have the highest impact because they ruminate (burp up methane). Beef also often has considerable deforestation associated with it, when land is cleared for feed production and grazing. Dairy products are lower impact than meats because it is more 'efficient' to keep the animal alive as a protein producer than to kill it. Of the meats, chickens are more efficient than the larger animals. They grow quickly, especially if you pack them full of antibiotics, and they don't waste much energy walking around much or keeping warm if you keep them in a crowded indoor space (see box). Note the land-use change resulting from animal feed for most meat and dairy. Note also how dramatically lower impact all the plant-based protein sources are.

All the staple carbohydrates are relatively low carbon, but rice comes out worst (see next question). Maize comes out best because corn is a particularly efficient photosynthesiser. Note

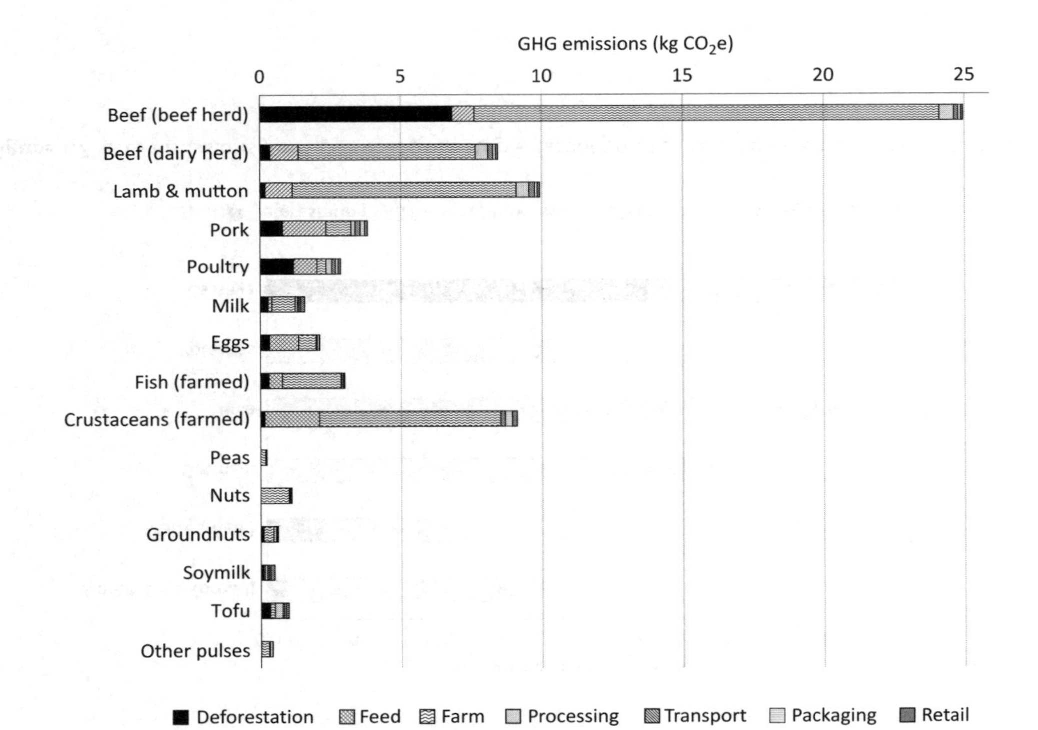

Figure 1.6 The GHG footprint of common protein sources per 50 grams of protein, broken down by supply chain stage.

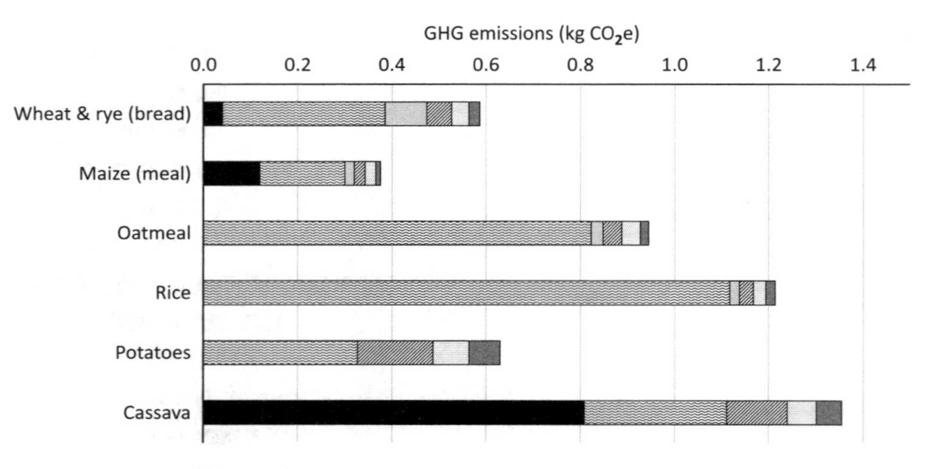

Figure 1.7 GHG emissions per 1,000 kcal from staple carbohydrates, broken down by supply chain stage.

the deforestation associated with cassava, much maize production and some wheat.

Is chicken the best meat?

Judging from its emissions, chicken might be thought of as the most 'environmentally friendly' meat, perhaps the best option for the caring omnivore. However, despite chicken's relatively glowing carbon credentials, poultry farming poses a whole host of other problems for both the local and global environment.

Broiler chickens (those that are raised specifically for their meat) have been bred to grow very quickly, and this leads to many welfare concerns including skeletal deformations and congenital heart defects. Stuffing as many birds as possible into a barn encourages the spread of disease, both viral and bacterial. Using plenty of antibiotics helps with the latter in the short term but contributes to increasing microbial resistance to drugs. So one of the trade-offs with this farming system is between the antibiotic resistance you encourage by using antibiotics against disease mutations versus the risk of human pandemics if you don't. Nasty for humans either way. Clearly the most ethical way to raise chickens is free-range, and a myriad of environmental impacts have emerged in recent years as demand and legislation for free-range birds has increased.

Most notably among these impacts, pollution. Chicken poo is full of nutrients, particularly phosphates. Raising hundreds or thousands of these birds outdoors results in a lot of poo, which can be washed off farmyards and pastures after rain and floods. The nutrients enrich local waterways and water supplies, triggering harmful algal blooms. Furthermore, chickens are often fed on soya, sourced from abroad, and associated with deforestation and other harmful farming practices.[22]

Should I go veggie or vegan?

Great idea! Lower consumption of meat and dairy is essential for the food supply, climate and biodiversity. But, none of us needs to go quite all the way unless we want to.

Some of the world's meat and dairy still comes from grass-fed animals. And some of that grass is grown on land that is not fit for crops. While it is true that some of that space might be better used for other environmental purposes, such as carbon capture and biodiversity, we should not dismiss the role that animals can play in turning nutrition that we can't digest into food that we can. Animals can add to the variety and health of our diets but there is an undeniable need for a reduction to perhaps half today's global average meat and dairy consumption. It means we will need to see much deeper than 50% cuts in meat consumption in most richer countries. It means reversing the current global trend towards more meat, which the United Nations Food and Agricultural Organization predicts will lead to a 23% increase in meat and dairy eaten per person by 2050.

Because food conversion rates are higher for dairy than for meat, while a move towards less meat and dairy is the most helpful dietary change, swapping meat for dairy is also helpful.

The first piece of really good news is that most meat eaters could make both these transitions in ways that enhance not only their health but also the variety in their diets. Just a personal view, but in a taste and texture contest, surely veggie haggis slaughters the traditional version? A second piece of good news is that, as I write, veganism is gathering momentum in many rich countries, providing perfect role models for the world's aspiring middle class.

Cutting consumption of meat and dairy also turns out to be the best place to start when it comes to cutting the carbon in our food (I'm using carbon as a shorthand for greenhouse gases). I've written about food greenhouse gas emissions extensively before, not least in my first book, *How Bad Are Bananas?*

The Carbon Footprint of Everything, which I now unashamedly plug along with a couple of academic papers on carbon and diet.[23] To summarise all of that, the top priority is to cut down on the ruminant animals: cows and sheep.

There is no need for extremism here, just moderation and a broadening of choice. If everyone who has that choice asks 'Shall we have meat, pulses, beans or eggs tonight?' in the same open-minded way that they might ask 'Shall we have pasta, potatoes or rice?', that shift alone could generate enough space in the whole food system to feed the 2050 population, while allowing more land to be devoted to biodiversity and, at the same time, if we wanted, liberating a bit more land for biofuel too (see page 49 and Are biofuels bonkers? on page 88).

If and when you do eat meat, as we have seen, some types are more impactful than others, but all have a bigger climate impact than their plant-based alternatives. We see a hierarchy of carbon footprints with pulses, grains and soya beans as the clear low carbon winners, dairy and poultry products as runners up, and red meats in worst place.

What can shops do about meat and dairy habits?

Make alternatives to meat and dairy delicious and tempting.

There has been a widespread myth that supermarkets are forced to sell what the customer wants to buy, and therefore have no control over what they put on the shelves. This is clearly nonsense. Supermarkets are highly practiced at steering you towards the product with the highest profit margin. For over a decade now I've been working with one food retailer that completely understands the influence they can have. They are not a company that sets out to be radical and they certainly experience no less commercial pressure than the larger supermarkets with whom they compete. But they understand that it is possible sometimes to make the most sustainable foods look the most attractive to buy. A lot of my work for them has been about encouraging them to do this more often. Sometimes they remind me of commercial realities that they think I may have forgotten. But every now and then we find a new area of improvement; pushing seasonal vegetables, making the meat alternatives look delicious, a Christmas brochure full of vegetarian options, a new range of local seasonal flowers to replace less sustainable imports, advice on the use of left-overs . . . There are still big areas I'd like to see more progress on, but overall the change I have seen has been significant and genuine.

The single biggest area for any supermarket to work on is the ratio of meat and dairy to alternatives in their sales mix. As we have seen, even within meat and dairy there is a hierarchy of impacts, with beef sitting clearly at the high impact end (and cheese not far behind).

What can restaurants do?

If a customer chooses a vegetarian or vegan dish, they do actually expect to enjoy it just as much as if they had chosen meat. Sounds too obvious to be worth writing down, but clearly this message has not yet been universally grasped. Restaurants should increase the range of vegetarian/vegan dishes and make them at least as delicious, tempting and inspiring as anything else they sell.

What can farmers and governments do?

There are many factors to take into account. Alongside nutritional output, biodiversity and climate change there are important questions including animal husbandry, livelihoods, communities and traditions. Where I live there are even arguments about sheep as tourist attractions. These perspectives need to be considered simultaneously with open-mindedness, transparency, respect for the evidence and respect for the people involved.

Some things are clear from the science. The world needs fewer farmed animals and much less crop-based animal feed. We need to be more careful with fertilisers and antibiotics. Some land is unfit for anything other than biodiversity or grazing, and when working out what to do with it, we should bear in mind that over-grazing can trash biodiversity as well as soil carbon. It also looks clear that the right animals grazing in the right way can enrich the soil, including its carbon storage.

It is also clear that doing the right thing requires more work, and therefore jobs, than is required to simply maximise output in the short term. It takes care, skill and effort to look after land properly. So, a sustainable food and land system offers a huge net livelihood opportunity. This must surely be good news for farmers and their communities.

Even though there is a place for famers to try to improve sustainability where they can whatever support they get, the right incentives and subsidies will clearly make this a lot easier. Farmers, scientists and governments need to work together sensibly to make this happen.

How could one crop save us over half a billion tonnes CO_2e?

Over 1% of the world's total greenhouse gas footprint could be saved by simple improvements to the way rice is usually grown.

This is a big untold story. It has nothing to do with the fossil fuel used by tractors, lorries, boats or any other parts of the supply chain. Specifically, what is needed is more judicious use of fertiliser and not flooding paddy fields.[24] Paddy field methane is about 6% of all greenhouse gas emissions from the food supply chain. I've seen photos of rivers in China that are bright green from so much excess fertiliser; the yield is probably actually suffering from over application. Sounds simple, but until the issue gets more traction it is hard to make progress. I have been on the lookout for a sustainable rice supplier for Booths, the UK supermarket chain, for a few years now. It is proving weirdly difficult. The Sustainable Rice Consortium looked promising, but it turns out that while they would like Booths to fund them, they can't actually point to any sustainable production anywhere. For the moment, rice is the most greenhouse gas intensive of all the staple carbohydrates, by more than a factor of two.

A final note for perspective: despite all the scope for improvement, a vegetarian rice dish still counts as a relatively sustainable meal.

What can *I* do?

For the moment, buy rice a bit less often than you might have, until such time as you find a more sustainable supply. When you find such a supply, let me know at Mike@TheresNoPlanetB.net. Tell your friends and your shops. Make sure people you talk to understand the issue.

What can shops do?

Find a sustainable rice source and market it as such.

What can farmers do?

Use fertiliser sparingly – which might save money too. Don't flood the paddy field. Then market your sustainability credentials to boost sales.

Is local food best?

Only sometimes. Transport is usually a small component of the carbon footprint of foods.

Travel is usually just a small part of the carbon footprint of food. In my latest study for Booths, transport was responsible for just 6% of the carbon footprint of all goods at the checkout.[25] The big greenhouse gas deal is in the farming (see What's the carbon footprint of agriculture? on page 25).

Food transport only really becomes a big problem when things get put on an aeroplane. UK examples of this can include grapes and berries from California, fresh tuna from the Indian Ocean, baby vegetables from Africa and, perhaps worst of all, asparagus all the way from Peru. (You can't eat flowers, but many of them also travel on planes so the same rule applies.)

By contrast, putting food on a boat, even from the other side of the world, can enable a relatively sustainable food supply. What turns out to be a fairly small transport energy demand comes in return for an important flow of nutrients from places with plenty of sun and fertile land to highly populated places that are unable to meet their own food needs. Nor are a few hundred road miles a disaster, although the fewer the better, especially when it comes to heavy stuff like beer. So, a pint from your local brewery probably beats any other alternative, unless, as is sometimes the case in the UK, it comes to you via a warehouse at the other end of the country. Local tomatoes grown in an energy-intensive hot house in winter could be many times less sustainable than the shipped alternative from a sunnier part of the world. (And in the case of flowers, hot housing them out of season is no better than putting them on a plane.)

There is no place for air-freighted food in the twenty-first century.

To summarise, come the sustainable world, there simply won't be any air-freighted food. In the meantime, you can help by avoiding it where you can and having done that, you can largely relax about food miles and perhaps just use that argument as one more reason to enjoy a local pint in preference to one from thousands of miles away.

To tell whether something has been on an aeroplane, check the country of origin and ask yourself whether it has the longevity to survive the journey by ship, train or lorry. Bananas, apples and oranges usually can survive, whereas strawberries, grapes and asparagus generally can't. If something is locally grown but out of season it will have to have been hot-housed, which can be just as bad as flying. A UK example would be Scottish strawberries in January.

(A quick plug is irresistible here: you can find much more on this in my first book, *How Bad Are Bananas?*)

Sadly, if you just look at things from a local and short-term point of view, crop monocultures and intensive cattle farming, dependent on fertiliser, pesticides and bucket loads of antibiotics (see page 21) can be fine, and even deliver the most profitable yields. So, biodiversity management gives us yet another example, as if it were needed, of the inadequacy of the free market to deal with Anthropocene challenges.

Well done for reading so much bad news, rather than turning away. Having faced some difficult realities it is now high time for us to look at what can be done with our food and land to sort things out.

Where does fish fit in?

The world catches or farms 80 million tonnes of fish per year. That is about 12 kg per person per year or 30 grams per person per day. This could just about be sustainable, with care.

The fishing industry ranges from village canoes to huge trawlers out at sea, manned by slaves and plundering the seas in whatever way is most profitable, refuelling and transferring their catch at sea to avoid all forms of governance.[26] Roughly half of all production is industrialised trawling and farming, while the other half is small-scale hand fishing. There is about 10 million tonnes of by-catch per year (4 grams per person per day) – that's the stuff that is caught by mistake and thrown back, probably dead, into the sea. Small fisheries currently provide an important source of essential nutrients (zinc, iron and calcium as well as protein[27]) in many poorer parts of the world. Access to this supply depends firstly upon the small-scale fisheries not being over-run by industrial trawling, and also upon the local fish not entering the global market, at which point the world's poor are unable to afford it. Thirdly, climate change is likely to alter migration patterns and habitat areas of some fish,[28] with serious consequences for some communities.

Fish stocks globally are under huge pressure. The Marine Stewardship Council (MSC) estimates that 90% of world fish stocks are currently fully or over exploited.[29] So, for all fish's relatively low carbon credentials, there is no scope for increasing total supply and perhaps the world should be cutting consumption. Even if not, those of us in richer parts of the world need to think carefully about who might not get to eat fish so that we can.

Is farming the solution? Sadly, farmed fish are just farmed animals that swim, and the moment you get into this you incur all the problems that are associated with most of the world's animal farming; fish feed is no more nutritionally efficient than giving human-edible food to animals; farmed fish are often plied with antibiotics and polluting chemicals; the overcrowding can be similar to that found in factory farms. While a sustainably caught wild fish might arguably be thought of as a sustainable nutritional food bonus, farmed fish cannot.

The MSC certifies sustainable fish brands but their credibility may not be as high as we could hope for. Here's the catch (haha). It turns out to be a 'for profit' organisation that makes, for example, £10 million from certifying a big fishery (see page 203) for my guidelines on how to work out who can be trusted). As I write this, the MSC looks to be on the verge of certifying fisheries that do pole and line fishing one day, and then use the same boats to trawl indiscriminately the next.[30]

When is a seabass not a seabass?

When it is a Patagonian toothfish – renamed into a Chilean Seabass for promotional purposes.

The price and popularity of fish seems to have little to do with taste or nutritional content, and everything to do with marketing. To give just one example, Patagonian toothfish was undesirable until a Californian fish merchant marketed it as the new-found delicacy of Chilean seabass in the late 1970s,[31] pushing the price to over £60/$85 per kilo. It isn't even a seabass! Sadly, the result of all the popularity is that stocks of this once-abundant fish, found in the deep (by which I mean anything from 300 m to over 3.5 km down) Southern Ocean, and able to grow to over 2 m in length and 100 kg in weight, are now threateningly low. And it takes 45 years to replace a 45-year-old fish. Over 80% of Chilean seabass, once Patagonian toothfish, is thought to come to us through unregulated fishing.

Another example of the many fish that have been renamed to make them sound more familiar to un-knowing consumers is the changing of the weird 'eitch' to the delicious 'Torbay sole'. Similarly, the gross-sounding 'slimehead' has become the exotic 'orange roughy'.

It is good that we can be persuaded to like anything that is available, but not if the ensuing fad leads to stock decimation.

How can we sustain our fish?

What can I do?

Here are six guidelines that I think seem sensible for non-vegetarians.

- *Treat fish as a treat.* The global average of 30 grams per person per day would only be sustainable with major improvements to existing fishing practices and transparency. Even then, most of us would still have to eat less in order to allow those who have to rely on it for their main nutrition to have what they need. So 30 grams per day would be two small fish meals per week or one larger one.
- *Find a fish monger who can talk to you convincingly* about where their fish comes from and how they know that slavery, by-catch and over-fishing are all minimised, and secondly can advise you on the sustainable options of the

day. Specifically, the Sustainable Food Trust[32] suggests questions along the following lines:

  - 'Can you suggest a fish for me to buy today? I want to try something different, which is sustainably and ethically sourced!'
  - 'Can you tell me about how and where it is farmed or caught?'
  - 'Why do you source from this particular farm or merchant?'
  - 'What are the environmental and ethical issues to consider with this particular type of fish?'
  - 'How do the seasons affect what fish I should be eating right now?'

- ***Be open to different species***, including obscure, unfamous varieties that you may never have heard of. These will probably also make your diet more interesting. If you can, buy from someone who can tell you how best to cook it.
- ***Don't let price or marketing count as evidence of quality***, because it probably isn't. But equally, be prepared to pay more for your ethical and sustainable treat.
- ***Take note of sustainability labels but treat them with caution.*** For example, 'dolphin friendly' is a red herring on any tuna of the skipjack variety, as it does not, and has never, swum alongside dolphins. The 'Pole and Line' label is probably still worth something. Sadly, there are no effective labels to let you know how much slavery you will be supporting.
- The MCS (and NOT to be confused with the MSC) produces a valuable and accessible ***Good Fish Guide***.[33]

What can shops do?

- Understand your supply chains. Source sustainably and let your customers know what you are doing. Take heed of sustainability guidelines from the MSC but go deeper. Avoid the naughtier brands, and don't be scared to boycott in the light of new findings, even if they are household names.

- Vary your fish stock in line with sustainable availability, and educate your customers towards a more interesting and wide-ranging fish taste. Let them know why you are doing this.
- Avoid air freight. If you do need fish from the other side of the world, properly frozen and put on a boat is the better option by far.
- Finally, help your customers to understand that fish is a valuable and limited resource. Make sure your sales people can give good answers to the customer questions listed above.

What can governments do?

- Ensure your own waters are sustainably fished. Easier said than done if there are fish pirates and boundary disputes, but this is the challenge.
- If fish is an important source of nutrition for your people, don't let it enter a global market until your own population can afford to buy at those prices.
- Police the industry, rooting out the slave trade as best you can.

What can fishermen do?

Is this too obvious to mention?

- Don't over-fish.
- Make sure everything you catch is eaten.
- Sell locally where you can.
- Don't run a slave ship.
- Do stick to the rules.

Now we move on to look at waste, the second biggest issue that jumps out of the global calorie and protein flow maps.

What food is wasted, where and how?

Out of 1,320 kcal wasted per person per day, 48% is cereals. That's enough calories to feed everyone in China and America. Nearly two thirds of all losses occur in harvest or just afterwards, in storage.

Table 1.1 The proportions of all calories wasted, by region, food type and stage in the journey from field to mouth. (Due to rounding errors, totals do not always equal the sum of the components.)

Regions / Waste stage	Harvest	Post-harvest	Processing	Distribution	Consumption	Total
Africa	4%	4%	1%	1%	<1%	**10%**
Americas	9%	2%	1%	1%	9%	**22%**
Asia	17%	21%	3%	7%	5%	**53%**
Europe	4%	3%	1%	1%	6%	**15%**
Global	**34%**	**30%**	**5%**	**10%**	**20%**	**100%**
Food groups / Waste stage	Harvest	Post-harvest	Processing	Distribution	Consumption	Total
Cereals	15%	17%	<1%	3%	13%	**48%**
Roots and tubers	3%	4%	<1%	1%	1%	**9%**
Oilseeds and pulses	13%	8%	<1%	1%	1%	**23%**
Fruits and vegetables	3%	1%	<1%	2%	2%	**9%**
Meat	<1%	<1%	2%	2%	2%	**5%**
Fish and seafood	<1%	<1%	1%	1%	<1%	**2%**
Milk	<1%	<1%	2%	1%	1%	**4%**
All foods	**34%**	**30%**	**5%**	**10%**	**20%**	**100%**

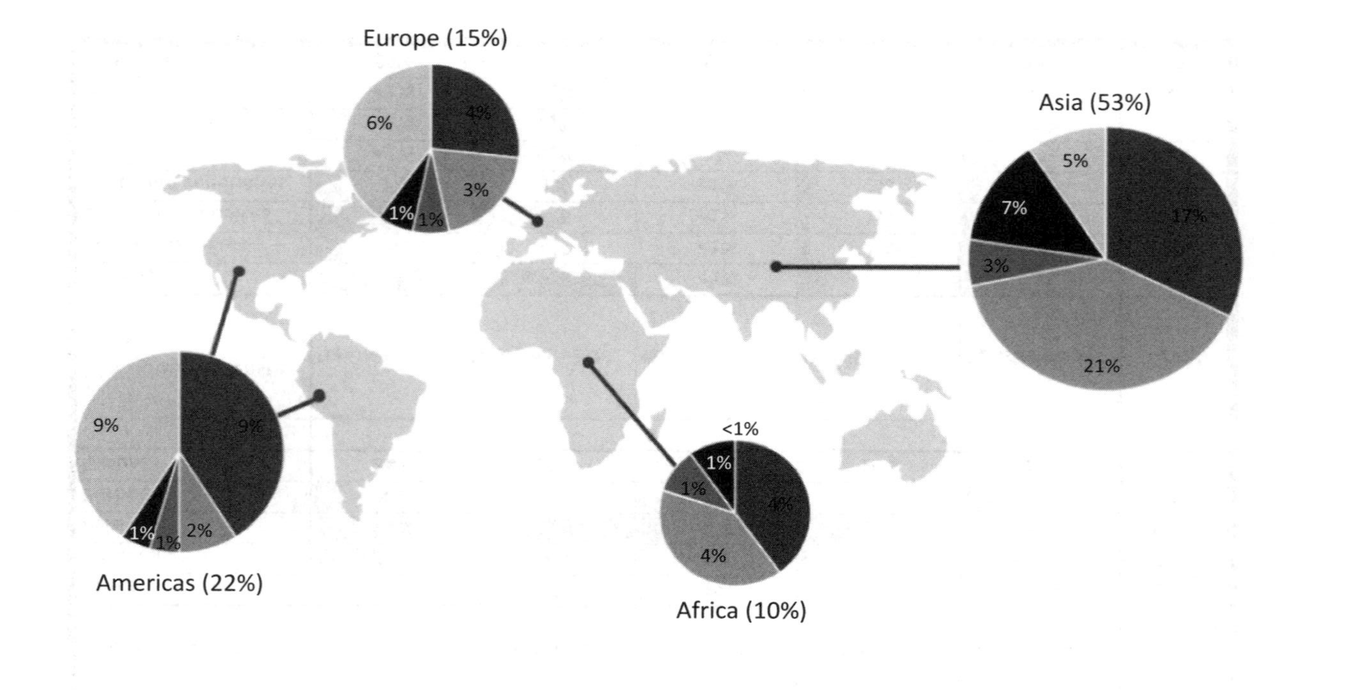

Figure 1.8 Where and how is food wasted (Oceania not included, but accounts for <1%)?

After dietary change away from eating too much meat and dairy, cutting waste is probably the next most important way to ensure there is enough nutrition to go round.

It is tempting to rage about waste everywhere we see it. This is fine as far as it goes but if we are serious about making improvements, we'd better have a closer look at how much is generated and where. Only then can we begin to prioritise. Waste stats are often quoted in tonnes, which doesn't help us because it gives a kilo of watermelon the same significance as a kilo of discarded beef or cheese. So I am going to talk about waste in terms of calories lost (protein matters a lot too, but the story turns out to be quite similar[34]).

Consumers account for 20% of all food waste, of which three quarters comes from the one quarter of the world's population living in Europe and the Americas. Even more seriously, but

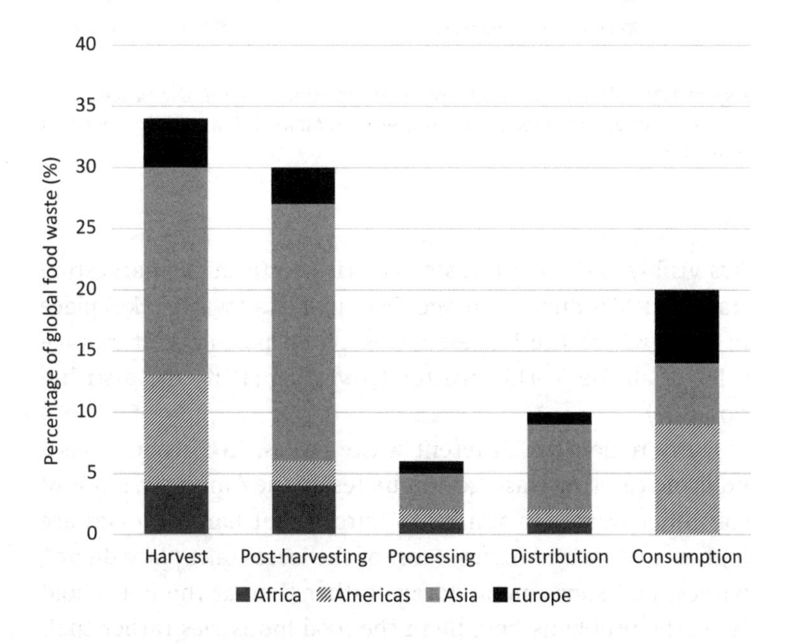

Figure 1.9 Global food waste by region and stsge in the process.

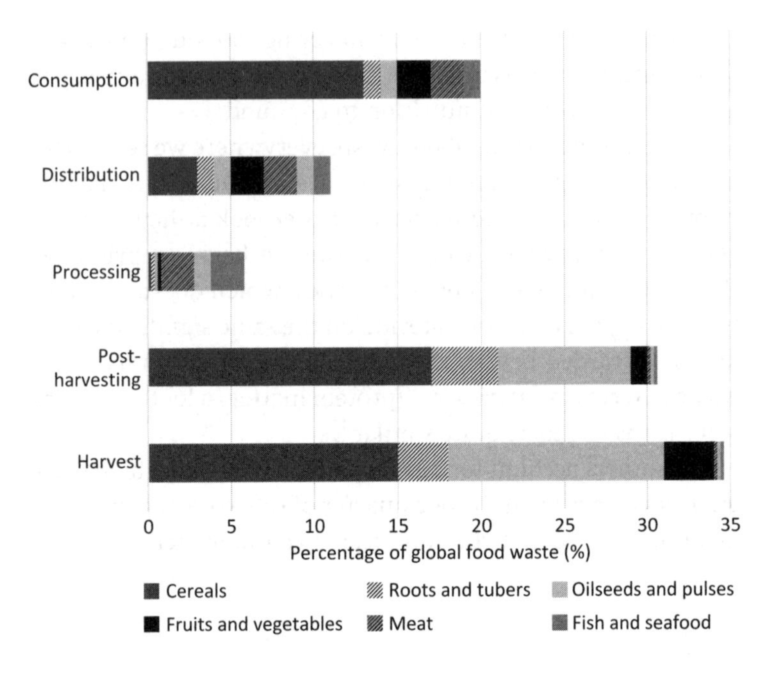

Figure 1.10 Global food waste by food type and stage in the process. Cereals account for 48% of all calories lost. Meat, fish and dairy together make up 9%.

less visibly, 34% of all wasted calories occur at the harvesting stage and 30% during storage. Over half of all waste takes place in Asia, where the biggest losses are in post-harvest storage (21% of all the world's waste), harvesting (17%) and distribution (7%).

Every region has different waste issues. In Europe, household and catering wastage dominates. In the Americas, levels of consumer waste are similar to Europe, but harvest losses are higher. In Asia and Africa, most of the losses take place during harvest and subsequent storage rather than at the household level; the problems here lie in the food industries rather than with careless consumers.

In terms of which foods get wasted, cereals account for 48% of all calories lost, while meat, fish and dairy together account for just 9%.

How can we cut the world's waste?

Cutting waste by half would add 20% to the world food supply.

In developing countries this mainly means more efficient harvesting and storage; in developed countries and among the world's rich, it is more to do with eating everything we buy. Most of what is required boils down to some relatively simple facilities in poorer countries and some cultural change in richer countries.

What can *I* do?

Eat what you buy. This sounds too obvious to mention but the developed world is spectacularly bad at it. This action is about looking in the fridge before deciding what to buy or what to have for dinner. Don't 'buy one-get one free' unless you know you will eat it. Learn to do great things with left-overs. The elimination of household and catering waste in just Europe and the Americas alone would add 10% to the world food supply.

What can restaurants do?

Help customers have only what they will eat on their plate. Options include self-service and portion size choices. And, of course, offer boxes and bags (recyclable ones) for left-overs.

How can shops help us cut waste?

The biggest thing a shop can do to help is to encourage its customers to cut waste. To do this also means being a good role model.

Supermarkets need to help people to buy only what they will eat. To some managers, this sounds like commercial suicide, but there is a lot of customer trust to be gained, as well as increased ability to live with yourself. Apart from skilful stock management and discounting, here are some of the things I have seen successfully tried at Booths, my semi-local supermarket chain:

- Buy one-get one free *next time*.
- Selling fruit and veg loose so people can get exactly the amount they require.
- Train the fresh counter staff to help single people buy single portions.
- Sell a good range of products for preserving left-overs.
- Promote menus for left-overs – especially after Christmas (or Thanksgiving).

Cutting their own waste further of course sets a good example, and customers are rightly upset when shops throw food away rather than giving it away.

As I write this, I am sharing the office with an intern, Sam, who does his local supermarket a favour most weeks, easing their conscience by raiding their bins for food. He periodically comes to work with a backpack full of random items that his own student network can't deal with. Between us all we do our best to make sure it all goes in to human mouths somehow. The office feels like a food bank at times, except that instead of providing for the genuinely needy, it often turns out to be more about middle-aged people like me eating more cake than is good for them. The difficulty for the supermarket is that by the time a product is so close to its end of life that it can't be sold, even for a tenth of its original price, it is very hard to give it away either. Sam brings in only low risk foods and they look fine too, so we all feel safe, but the supermarket has to be more careful still. The result is that there is not time to distribute it out via real food banks. A way round this might be to rapidly freeze stuff just before the use-by date, but for that to work the supermarket needs to care enough that it will stand both the extra cost and the extra hassle, compared to throwing it in the bin.

Why don't supermarkets care more about their waste?

They already care a great deal because it is so expensive for them. A well-run supermarket will lose less than 0.5% of its food, whereas the average European household wastes nearly a quarter of food that it buys.

Some of the most headline-hitting sources of waste actually turn out to be some of the least significant. Retailers in Europe and America, for example, are already highly incentivised not to throw away food because it is so expensive. For the same reasons, manufacturing waste levels are also fairly low. So while it is right that supermarkets shouldn't throw stuff in the bin that they could give to food banks, and while it is important that they are seen to be role models, we also need to recognise that this is a small part of the total waste problem. Finally, for perspective, remember that waste in itself is a smaller issue than the rising level of meat and dairy in the global diet.

When food can't be sold or eaten, what should be done with it?

Feed all human food to humans whenever you can. Avoid landfill. Be careful with garden compost. Don't get too excited about any other options.

Having done everything possible to get all its food to the check-out, a supermarket is still bound to end up with some that it can't sell. The chart below is based on some recent research we did at Lancaster University to help UK retailers understand their options.[35] It describes how good each disposal pathway is for mitigating the waste from a greenhouse gas perspective: 100% indicates that the problem has been completely solved, 0% represents not having mitigated the problem at all, and negative numbers indicate disposal pathways that actually make things worse by creating yet more emissions.

Emissions mitigation (%)	Bread	Cheese	Fruit and veg	Fish	Meat	Average food
Donated	100%	100%	100%	100%	100%	100%
Fed to animals	24%	7%	1%	41%	5%	6%
Anaerobic digestion	20%	4%	5%	19%	4%	6%
Composted	3%	1%	−1%	5%	1%	1%
Incinerated	11%	2%	−2%	1%	1%	1%
Landfill (gas captured for electricity)[36]	−44%	−7%	−12%	−26%	−7%	−10%
Landfill (gas captured but flared)	−61%	−10%	−16%	−36%	10%	−14%
Landfill (no gas capture)	−227%	−37%	−61%	−136%	−36%	−53%

Figure 1.11 Savings in greenhouse gas footprint of foods resulting from different disposal options. All solutions are rubbish, except for donating it for human consumption.

The first thing to see is that finding a way of donating the food so that it is eaten by humans is the perfect solution, and the only one that should be thought of as satisfactory. You incur a little bit of extra transport, getting it to a food bank, say, but the impact of that turns out to be insignificant.

All the landfill options make things worse by generating methane, a very powerful greenhouse gas that is impossible to completely capture. Some landfill sites leak more methane than others.

In the middle of the range are a bunch of solutions that sound good but whose success is largely limited to not making things worse. Bread and fish typically have low carbon footprints compared to their calorific content, and this means there is a bit more benefit in feeding them to animals, or burning or anaerobically digesting them to generate electricity.

What can I do?

Households can take away many of the same messages. Give your food to a friend or neighbour if you can't get through it yourself. Whether the stuff you throw in the bin goes to landfill or not will probably depend on your local authority. Where I live the food in general waste 'grey bins' ends up being turned into fuel for incinerators.

One note of caution is that your garden compost has to be turned over often enough that it rots aerobically (i.e. with access to oxygen), rather than anaerobically – which is roughly like having the worst kind of landfill site in your own back garden belching out methane.

After animals and waste, third on the list of big food issues that leap out of the global food nutrient flow maps comes biofuel.

How much food goes to biofuel?

The answer is 810 kcal per person per day. That's the same as a 10" margarita pizza every day for everyone in the world.[37] It creates enough fuel for everyone to drive just half a mile in a traditional, oil-burning car.

After animal feed and waste, biofuel looks like the third biggest cause of loss to the human food supply. If we are being precise about it, the figure here is for all 'non-food uses' and these also include cosmetics, pharmaceuticals, paints, plastics and all sorts of stuff. But it is mainly biofuel. We will see later that biofuels are mainly bonkers (page 88). Enough wheat to provide the daily calorific requirement for one person for a day is only enough to power a small petrol car, such as my Citroen C1, for one and a half miles. If biofuel for cars became popular, it would lead to a lot of hunger. We need to watch this like crazy as we move to the low carbon world. To spell out the threat, a carbon price that is high enough to see the fossil fuel staying in the ground would mean that in a free market, wheat would often be more profitably directed to biofuel than to human mouths.

How many farmers does the world need?

More than the 1.3 billion that we have now. The good news is that the world has no shortage of people.

The number of people working on our land today is falling. But by 2050 there will be about a billion more of us of working age. As the population climbs to 9, 10 and 11 billion, keeping everyone busy could well become increasingly challenging. Yet because people cost money to employ, the free market ensures that much of our agriculture is geared to minimising the number of farmers per square mile. For all our technology, it still takes personal care and attention to grow food productively, environmentally and even beautifully – and doing this must surely be one of the most inherently positive ways that a person can spend their time on the planet. So we need top-down interventions to encourage more people to work on our land. Neoliberalism's free market, as we will see throughout the book, turns out to be unable to deliver what we need.

Governments: since the free market can't look after the land, in large part it will be up to you to get things working properly. You need to incentivise the right things and set up subsidy programmes to get a sustainable agricultural system in place.

How can new technologies help feed the world?

As we've seen, with enough societal change and waste cutting, and without adverse effects from climate change, no new technology would be needed. But if sensitively applied, it can make life conciderably easier.

In other words, unless climate change severely reduces land productivity, it is not true to say that we *need* new technologies in order to get by but it is also not true to say that technology alone will solve the problem. And we will also see in a few pages' time that there can be no place for any technology that gets in the way of a shift towards a more biodiverse agricultural system. On the other hand, if the climate crisis reduces yields or the population grows higher than the 9.7 billion predicted in 2050 then further measures become necessary.

A range of emerging technologies from genetic modification to lab meat or using solar power for irrigation are all on the cards. All of them need applying with sensitivity, without which some of the possibilities on my list conjure up nightmare images. They might even allow us to deal with a global refusal to cut meat and dairy production. The essential point to grasp about emerging technologies is that while they stand to be helpful if used in ways that are sympathetic to both people and planet, they will not in themselves create a world in which everyone has enough, or in which biodiversity is preserved or in which our essential relationship with nature is restored. However, key helpful technologies include:

- Indoor plant farming: There is a very unromantic reality that the most efficient way to grow food is going to be to house it in special tower blocks for plants, sometimes referred to as vertical farms, with lighting powered by solar panels and every nutritional input carefully optimised with high-tech monitoring and the latest algorithms.
- Lab meat: While this may be no more appealing a concept than indoor plant farming, this could offer a considerable improvement on most of today's meat industry.[38] Potentially tasty, efficient and humane.
- Water technologies: Ways of growing more with less, using green energy for irrigation and desalination. In short: food from the deserts. Graphene promises a huge advance in desalination efficiency while the solar revolution provides the power.
- The development of rice strands that are capable of a more efficient type of photosynthesis, as deployed by maize. (Sometimes called C4 rather than C3 photosynthesis.) The Gates Foundation is pumping £14 million into this.[39]
- Genetic modification: Carefully applied and freely available, this could help with higher yields, better

nutritional content and lower greenhouse gas emissions, less water consumption and better climatic resistance.
- Waste reduction apps are emerging to connect food that needs eating quickly with people who could put it to good use.

Even simpler than new high-tech solutions is the propagation of well-established best practice such as judicious use of fertiliser and reduction in paddy field flooding.

UPDATE. Over the past 12 months I have been getting my head around factory production of both carbohydrate and protein. The reality is that this stands to be far more efficient than plant-based food.[40] Solar panels can convert roughly 20% of the sun's energy to electricity and a pilot factory in Finland claims to be able to store electrical energy as carbohydrate with 20% efficiency. Overall, that delivers carbohydrate from sunlight with an incredible 4% efficiency. This is perhaps 50 times more energy efficient than growing wheat. Similar efficiency gains look possible with protein production. If this is the case, every hectare of land devoted to solar panels for food production could enable a further 50 hectares to be used for biodiversity. To me, it isn't an immediately appealing food solution, but if it can help us to feed everyone while improving biodiversity and dealing with the climate emergency, perhaps it is worth it. Of course it puts even more pressure on resources required for solar power. And we should remember that if it is not combined with a more sensitive approach to the planet, all we will end up with is factory-created food, *alongside* a climate disaster and a biodiversity collapse.

How can we produce enough food for 9.7 billion of us in 2050?

As we have seen, the priorities are (1) to reduce human-edible food being fed to animals, (2) to cut waste, (3) to keep biofuels in check and (4) the sensitive application of new technologies.

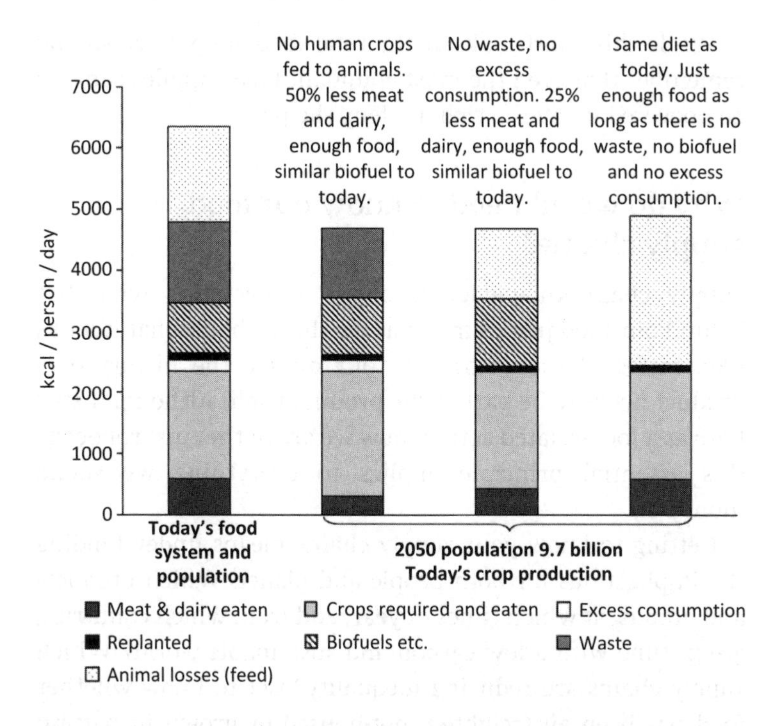

Figure 1.12 Food scenarios for 2050. In 2050 we can happily feed 9.7 billion people if we halve the waste and cut by 80% the amount of human-edible food we feed to animals. This will lead to the amount of meat and dairy per person dropping to about half today's global average. On the other hand, if meat and dairy consumption per head stays the same we will be in deficit unless there is no waste, biofuel and excess consumption.

The chart shows some scenarios to 2050, based on exactly the same crop production that we have today.[41] It shows what happens when the population rises to 9.7 billion (as predicted for 2050[42]), and how we could get by under different levels of meat and dairy consumption and waste. In each scenario, note that the biofuels wedge does not have to be used for that purpose – it can be read as a measure of spare capacity in the system to liberate land for other environmental purposes, including biodiversity and carbon sequestration.

While this chart looks at the total food supply, we should remember that even the most abundant food supplies can still be hoovered up by the rich, to deny the poor.

Why do we all need to know our food supply chains?

Supply chain knowledge is about appreciating what lies behind our food just as much as we think about what it looks like, tastes like and costs. In our minds, the history of a product needs to be part of the product itself. Although listed here as a food-related action, now we are in the Anthropocene this essential principle applies to everything we spend money on.

Getting to know your supply chains means understanding the implications for both people and planet. Which products and brands, at which times of year, and from which countries, are in tune with a low carbon and sustainable world? Which supply chains are reducing inequality? Get to know whether food has been air-freighted, hot-housed or grown in natural sunlight. If the information is hard to find, ask for it. If you are a restaurant or business owner, share your actions and your supply chain knowledge with your customers.

As well as a broad understanding, supermarket food buyers need to have a detailed understanding of the issues in their area, and their managers need to make this possible. The specific issues are different for every type of product, so buyers need to consider it central to their work. For some it will be about working conditions and pay. For others it is all about fertiliser use. Or deforestation. Or air freight. Or meat alternatives. Or all of these and more.

For food professionals and everyday shoppers alike, everything we spend money on is an investment into one type of future or another (see page 159). We all need to push our money into the supply chains that we want to see thriving. Ask where, ultimately, will the money you spend end up?

Who will get richer, and what will they in turn invest in? Buying food is a powerful act.

What investments are needed into food land and sea?

We need investment in schemes that keep our forests and grow food sustainably. And we need research into the agricultural practices that can put carbon back into the ground, and more generally on the soil and biodiversity implications of different agricultural practices.

What we don't particularly need is research and development of techniques for increasing yields at the expense of biodiversity.

Many of the required improvements should not require multi-billion dollar investments. The single most important change will be an amazingly simple dietary shift towards less meat and dairy consumption, with a particular focus on reducing beef. This will markedly reduce greenhouse gases, improve the nutritional output of our land and, by relieving land pressure, ought to be pivotal in stemming deforestation. The net infrastructure investment requirement should be nothing or perhaps even less than that! We also need to cut waste throughout the food chain but, here again, the infrastructure requirements are not vast.

However, there are two critical areas for which investment is required. The first is research. We don't yet know nearly enough about the impact of different arable practices on the environment and, in particular, what farming systems store or release carbon and in what quantities. Research is needed into how to grow efficiently while encouraging biodiversity. There are promising manufactured alternatives to meat that need looking into further. We also need to understand how land can be used to create the liquid hydrocarbons that we will almost certainly need if we are to continue with aviation in the low carbon world.

The second critical investment area is farmers. We need to understand that the best ways of dealing with our land are not the cheapest. To do a really good job of producing food, cutting emissions and promoting biodiversity requires care and attention. It requires plenty of people. The good news is that we have more of this resource than ever before and will soon have at least another couple of billion more still. For the past couple of centuries we have been looking to minimise the number of people working on the land. This is crazy given the abundance of person power. We should be looking to employ *more* people to do a *better* and more careful job of looking after our land and growing our food. We need to invest in farmers and subsidise them to do the right things. The money for this can be made available by ceasing the deeply unhelpful subsidy of fossil fuels, divesting from them and better still, applying serious taxes.

Food action summary: What can *I* do and what can be done?

At the global level, here are the five things that will help most:

- Change the dietary trend from more meat and dairy to less.
- Create limits to first- and second-generation biofuel[43] (to take pressure off the agricultural system and allow development of agricultural practices that support biodiversity).
- Improve targeting and efficiency of fertiliser, pesticide and water.
- Eliminate phosphates from detergents.
- Greater establishment of protected areas in land, sea and freshwater.[44]

At the personal level, here are the simple things that anyone can do:

- Buy and eat food in ways that enable a biodiverse agricultural system. Once again this takes us to lower consumption of meat and dairy, especially less beef and lamb, less waste, keep your fish consumption modest, and always from sustainable sources (see pages 38–39).

- Get to know your supply chains and buy food from the ones you like. In this way minimise the carbon, antibiotics, deforestation and slavery embodied in your meal. Maximise the biodiversity and quality of employment that lies behind every mouthful (see Investment, page 159).

2 MORE ON CLIMATE AND ENVIRONMENT

Our tour of the food system has already taken us into some big environmental challenges. But before moving on to look at our energy supply we need to look at the climate emergency more broadly, along with a couple of other huge issues that get a quick nod here but really deserve whole books on their own.

This section may seem full of bad news but stick with it. I have kept it short, but we have to look these realities in the eye so that when we move on to the opportunities and solutions that come later in the book, we do so in full view of what is required.

What are the 14 things that every politician needs to know about the climate emergency?

By 'needs to know', I mean that anyone who doesn't get all 14 of these things is unfit for political office. In the Appendix I flesh these out in more detail.

(1) **Current science tells us that a global temperature rise of 2 °C looks very risky but 1.5 °C much less so.** The Paris Agreements said this. Every significant country in the world agreed. Since then Trump has rowed back but everyone else has held firm.

(2) **The temperature rise we experience will be roughly proportional to the total amount of carbon we have ever burned.** This gives us a 'cumulative carbon budget', most of which we have already spent.

(3) **Emissions of carbon dioxide, the most important greenhouse gas, have grown exponentially for 160 years.** There are always ups and downs between individual

years and there was a bit of a dent for the Great Depression and the world wars, followed by bounce back that can be explained in various ways. But these are just noise against a remarkably steady long-term trend of 1.8% growth per year.

(4) **We have not yet dented that carbon curve.** Even taking the most recent data into account, there is little or no evidence that the carbon curve has even twitched in response to any of the talk and action on climate change so far. Gulp! (Facing this reality gives us a much better chance of dealing with the problem. This stark observation tells us a lot about the nature of the solution that we need to put into place.)

(5) **At the current rate of carbon emissions the remaining viable carbon budget for both 1.5 and 2 °C is dwindling quickly – despite some recent good news from the carbon modellers.** As things stand, we look set to overshoot the 1.5 °C budget sometime between 2030 and 2040.

(6) **It takes a long time to put the brakes on.** The temperature won't stop rising until net emissions are zero.

(7) **Almost all the fuel that gets dug up gets burned – so it has to stay in the ground.**

(8) **Because of rebound effects, which are often ignored, glossed over or not fully understood, some of the key actions that many people assume will help us haven't helped at all – and on their own, they never will.** This includes just about every new technology and efficiency improvement.

(9) **Growing renewables, while essential, won't be enough to deal with the climate emergency.** Precisely because of rebound effects and the permanent human appetite for more energy.

(10) **So, we urgently need a working global agreement to leave the fuel in the ground.** Piecemeal actions get

absorbed at the global system level by companies pushing emissions down the supply chains, countries off-shoring their carbon, and the myriad of ways in which the emissions simply migrate to elsewhere in the global system.

(11) **We need to manage other greenhouse gases too** (see Chapter 1).

(12) **Extracting and burning fossil fuel has to become too expensive, illegal or both** unless you can think of another possibility for a global constraint.

(13) **Such a deal will need to work for everyone.** In theory it might be possible to force some stakeholders to put up with an arrangement that sends them into poverty, but on the whole the whole world will have to sign up and help make it work. It doesn't matter how hard that seems, facing the challenge is an essential first step in meeting it.

(14) **We will also need to take carbon back out of the atmosphere – even though it is unclear whether we yet properly know how to.** This is quite simply because we are exposed to so much risk already through our failure to take effective action so far.

All 14 points might require a bit of digestion. Feel free to pause here and reflect. There is more detail on each one in the Appendix.

If only the climate emergency was the only environmental consideration. Let's check out some of the other giant challenges, not all of which get enough air time.

What are the biodiversity stats? And why do they matter?

On both land and sea, we know we are heading for trouble. The current mix of deforestation, ploughing up land for monocultures, over-grazing, over-fishing and the belching out of innumerable toxins, plastics and other pollutants has got to change. But just how fast, and how much do we really need to care?

*** **Warning: Well-adjusted readers may find the next paragraph frightening. I'd like to skip over it, but we can't.** ***

Nobody knows how many different species there are nor how fast we are wiping them out, but we do have estimates. There are probably between 5 and 10 million different species if you include all types of plants, animals, fungi and everything else, right down to single cell organisms.[1] We know that the biggest land-based animals were gone by about 10,000 years ago and it looks as if we are currently losing the remaining species at somewhere between 0.01% and 0.1% per year.[2] So if there are 10 million species that would mean between 1,000 and 10,000 lost per year. Excluding the microscopic stuff, in 2017 there were an estimated 25,000 species listed as 'threatened', up from 11,000 at the turn of the millennium.[3] And among the ones that are left, populations are falling through the floor. A WWF survey of nearly 4,000 species found a massive 58% reduction since 1970,[4] while a study of flying insects found a frightening 75% decline in just 27 years.[5] Gulp!

If you were to take life from a purely functional and human-centric point of view (which I don't), it would not be straightforward to prove the absolute necessity of all today's biodiversity. It could possibly be tempting to think that fewer flies would make life more comfortable for humans, and the numbers of tigers and polar bears might even feel irrelevant to our everyday lives. Perhaps the ecosystem could survive being trimmed down a good bit further, and still supply all humankind's needs. What a desperately sad way of looking at the world. But if we were to adopt it, the easiest thing to do might be just to take the risk, right? However, careful assessment of the evidence makes it very clear that those risks would be enormous. At the very least we can expect falls in yields of crops, animal and fish as well as loss of resilience against disease.[6] And what if we suddenly realised we'd gone a step too far? There is no going back from extinction.

What is ocean acidification and why does it matter?

Caused by CO_2 and potentially just as nasty as climate change.

Described by Jane Lubchenco, former head of the National Oceanic and Atmospheric Administration, as global warming's equally evil twin,[7] this oddly gets less than 5% of the coverage of climate change. The basic story is that CO_2 from burning fossil fuel finds its way into the ocean and the resulting acidification reduces the ability of sea life to produce shells and skeletons.[8] Any species that takes a hit passes on the pain to all those who like to eat it. Thomas Lovejoy (former chief biodiversity advisor to the World Bank) described it as 'like pulling the rug out from under the marine food chain'[9] and the potential result is catastrophic collapse of marine life. Once this happens it will almost certainly be incredibly difficult to reverse. Our loss of sea food would be just one of the enormous consequences.

Why doesn't this get the air time that the climate emergency enjoys? Probably because everyone is so exhausted debating climate change that their energy has run out. And perhaps because the impacts might seem even more abstract than the idea of climate change, which at least includes things we can imagine like floods and ice shelves collapsing. We get exhausted with bad news and problems. Luckily the solution to this one pops out of the hat as we deal with the climate emergency, so its effect on our thinking can mainly be to double the motivation to get off our butts.

What's to be done and what can I do?

Just the same as for the climate emergency, at the individual level, the action is to cut our carbon footprint and do everything else in our power to create the cultural and political conditions under which the world can leave the fuel in the ground. Policy actions are also the same as for the CO_2 part of reducing global greenhouse gas emissions, but it is important to bear in mind that for carbon capture and storage, it is essential that any underwater stored carbon is not allowed to leak back into the ocean.

How much plastic is there in the world?

An estimated 9 billion tonnes has been produced so far.[10] Of this, 5.4 billion tonnes has been chucked into landfill or scattered onto land and sea. If all of it were cling film it would be more than enough to wrap the whole planet.[11]

The world is now producing over 400 million tonnes of the stuff every year. Of all the plastic ever made less than a third is still in use, less than a tenth has been burned, and only 7% has

Figure 2.1 If all the world's discarded plastic were cling film, it would be more than enough to wrap the whole planet.

been recycled. Some 60% hangs around as rubbish. When it gets landfilled, at least you can argue that the carbon is going back into the ground where it came from and where it belonged all along. From a climate perspective this could be the best end point. An estimated 4–12 million tonnes per year end up in the sea,[12] where it turns up on the world's remotest beaches, on the ocean floor and in the stomachs of birds. It sometimes even makes its way back into our food chain where it can be found, for example, in a third of the UK's caught fish.[13] One of the most depressing aspects of humankind's blind destruction of the natural world has to be the realisation that these tiny bits of plastic will be there forever, more or less. Thousands of years from now, a microscope on a pristine beach will still find the little grains of human-made multicoloured sand that we have unthinkingly laid down over just the last few decades. No amount of cleaning will sort that out.[14] And the

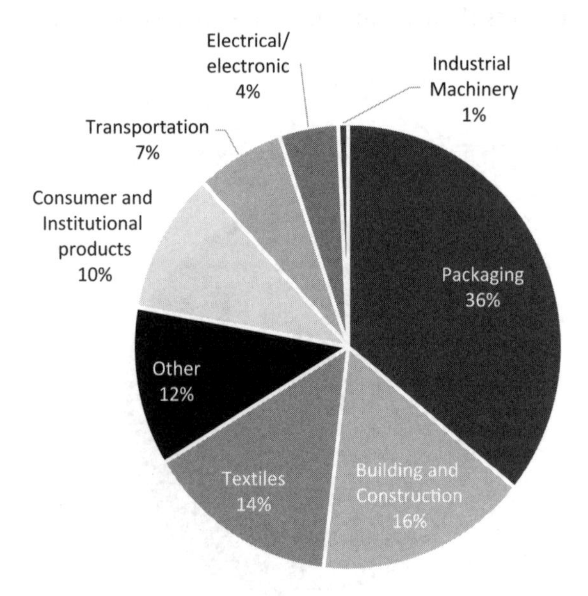

Figure 2.2 Global plastic production according to industrial use sector in 2015 (%). Over one third of plastic produced in 2015 was used for packaging.

amount in the ocean looks set to triple over the coming decade from around 50 million tonnes to around 150 million.[15]

Over a third of all our plastic is used for disposable packaging. To look on the plus side, recycling rates have gone from an all-time average of just 9% to nearly 20% today. But to look on the negative side, this is still pitiful and the growth in recycling rates has been dwarfed by the overall growth in plastic production, so that the amount getting chucked out every year is going up not down.[16]

Is fossil fuel better burned or turned into plastic?

Hobson's choice! Ideally it would have stayed in the ground. Landfill, for all its other problems, at least puts it back there.

If just one year's worth of today's oil production were turned into plastic it would almost double the total amount that there is in the world.[17] This sounds like a crazy notion, but already there are signs of oil companies looking to sell more of their wares to plastics factories, as the world slowly gets to grips with the need to not burn fossil fuels. However serious the climate emergency might be, shifting from CO_2 to plastic waste is even worse. So if you know someone that is proposing this business strategy, please call it out for what it is – poison.[18]

3 ENERGY

The big picture on energy now and in the future. A tour of energy sources, the practicalities of transition, and the deeper underlying challenge for humans.

Having looked at food, climate and environment, we turn now to the rest of our energy system, with which it is inescapably intertwined. It is our energy supply that gives us the ability to change our planet for good and bad. While a significant fraction of this supply is still consumed by mouth and another few per cent comes from biofuel the majority of both our energy and our greenhouse gas emissions come from fossil fuels.

Firstly, we will stand right back to get the big perspective; sketching out how much energy we use, where it comes from and what we do with it. Then we are going to look at what can be done going forwards. Most urgently, of course, the 'intensive care' side of the energy challenge is about dealing with the climate emergency. At the technical level, we will find encouragingly feasible solutions. But we will also have to debunk some of the popular ideas that won't work, one or two of which are deeply dangerous.

We will move on to look at some of the underlying dynamics of energy growth that are totally essential to understand. These are still not grasped or are glossed over by a frightening majority of policy makers, rendering them inadequate in their roles. Getting clear on these basics will be hugely helpful in looking at both intensive care and long-term health for people and planet.

Here goes!

How much do we use?

Humans use about one seven-thousandth of the energy that hits the Earth's land area. The average person uses 59 kWh

per day.[1] That is equivalent to about 6 litres of petrol, or enough to drive a fairly efficient petrol car for about 70 miles.

Alternatively, if it comes in the form of electricity, the average person's daily energy use could power a similarly sized electric car for about 280 miles or we could each have a toaster *and* a kettle on permanently; 59 kWh of jet fuel is enough for perhaps 100 passenger flight miles. In the form of human food, it is enough to give 22 people all the calories they need for a day.

My total figure includes all the wood that is burned for heat as well as the food that we eat. It is important to get the latter into the energy equation because, as we have seen, the food and non-food energy systems are increasingly intertwined through their use of land.

Of course, we don't all use the same amount of energy. The average European uses almost twice as much per year as the global average, the average American uses nearly four times as much, while the average African uses only about a fifth. Someone asleep in an unheated house is getting through only about 3% of the global average[2] while anyone travelling alone on a private jet will be running at about 1,000 times the average.

How has our use changed over time?

It has always been going up. And the *rate* of growth has risen too. We use more than three times as much energy today as we did 50 years ago.

Almost every year we have used more energy than we did the year before. There may be short-term blips, but the general trend has been going on at least since the Egyptians were using human (slave) power to build pyramids, and probably for a long time before that. The way it works is that the more energy we have, the more we can use for getting hold of yet more energy and for inventing more efficient and different ways of both using and acquiring it. Energy growth, innovation and efficiency improvements have always gone hand in hand. They have been racing along together as a team, gathering speed.

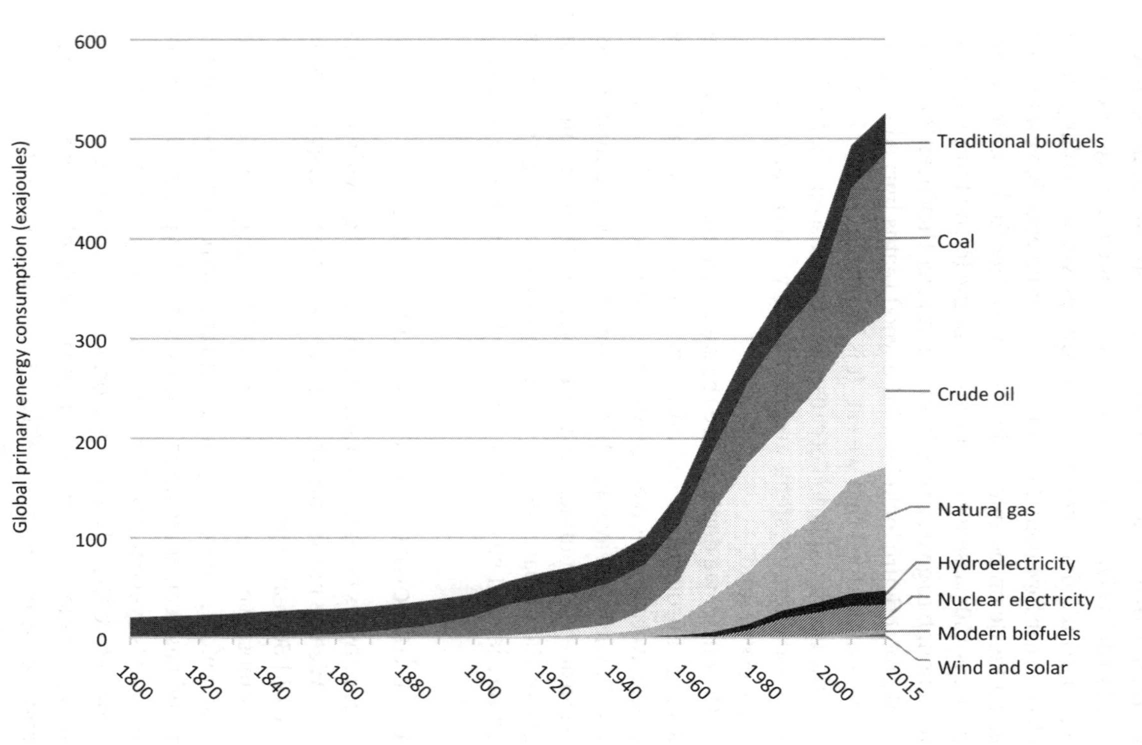

Figure 3.1 Global primary energy consumption, 1800–2015. It shows the rise of human energy use since 1800, with new sources augmenting rather than replacing the old.

I mentioned in the introduction that it is this rising energy use that has landed us, recently and accidentally, in the Anthropocene – the era in which we are big and powerful compared to our fragile planet, and in which we suddenly need to go about life in a completely different way if we don't want things to go terribly wrong.

Over time, the growth rate has risen steadily too. Centuries ago it must have averaged a small fraction of a per cent per year. Between 200 years ago and 50 years ago, it averaged about 1% and over the past 50 years it has averaged a massive 2.4% per year.

Every now and then throughout history, new energy sources have arrived on the scene; wood, then coal, then oil, then gas, and in small ways, hydro, nuclear, wind and solar. A quick look at the graph shows us that the arrival of a new energy source has not generally led to the decline in other sources. We have used the new energy source to augment rather than replace the old. The exception is a small drop in the use of wood fuel. Now, with all eyes turned to renewables, we need to be wary of this simple observation.

(If you zoom in to very recent years there is evidence of a dip in the growth trend. The average over the past 10 years has been around 1.6% and over the past 5 years as low as 1.3%. According to BP, 2016 saw only 1.0% growth. Some people get excited about this, but I treat it with caution for reasons that I'll explain in a few pages' time when we come to look at future energy use. For now, let's just say that there are always ups and downs in the short term.)

What do we use it *for?*

Around 5% is food used to power human bodies, and 38% is spent on transporting people and stuff on journeys of every scale. The rest is fairly evenly split between domestic and business use.

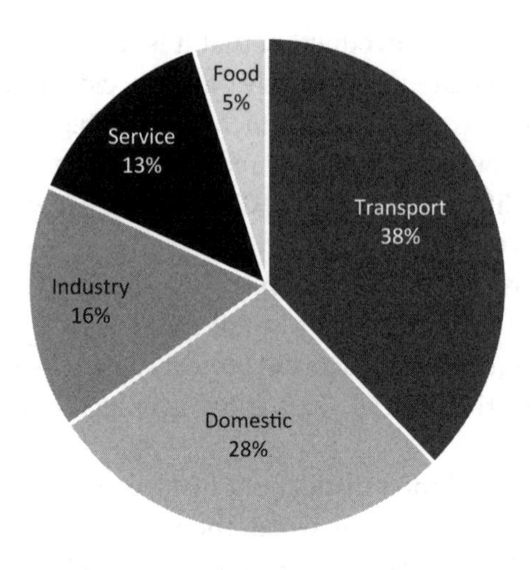

Figure 3.2 UK energy by end use.

Domestic use, at 28%, also includes cooking food, keeping us warm and, increasingly, for keeping us cool. A good chunk of the 16% industrial use is also devoted to the production and preparation of food, the provision of transport infrastructure and vehicles, and the creation and maintenance of homes, including all their cooking and heating apparatus. So one way and another, eating, getting around and keeping warm still account for the vast majority of human energy use, just as it has done through the millennia. The figures are based on UK statistics.[3]

Where do we get it all from?

Eighty three per cent comes from fossil fuels. Nuclear energy provides less than 2%. Renewable energy sources make up nearly 4%, of which two thirds is hydroelectric.[4]

Apart from the thin slice which comes from nuclear fuel and a tiny slither of tidal power (which you could say comes from the moon), our energy supplies all originate from the sun. Most of it was captured by plants long ago and laid down as the fossil fuel

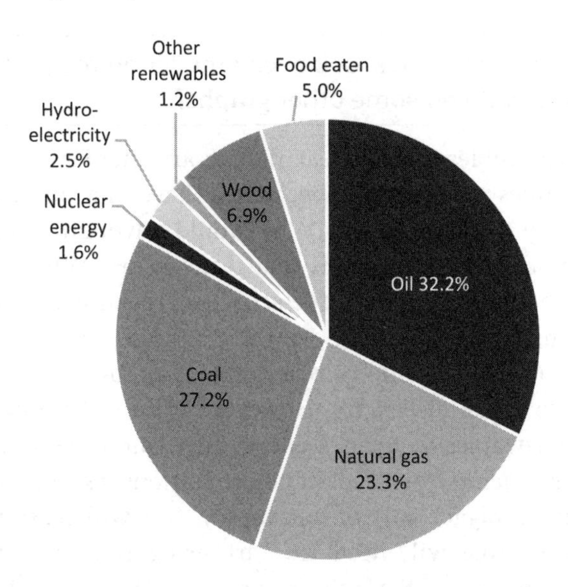

Figure 3.3 Humankind's energy supply of 18.6 TW in 2017, including food eaten. That is 2.5 kW per person (or 59 kilowatt hours per person per day).

store that we are now working our way through at ever-increasing speed. A sixth of today's energy supply comes from modern sunlight whether it is via plants, wind, rain or solar panels. Through photosynthesis, plants can typically capture between 1% and 2% of the sunlight that lands on them, and we get to use some of that energy as food, wood and modern biofuels.

A photovoltaic panel can do a lot better than a plant at capturing the sun's energy. Even a cheap one can harness 16% while a more efficient (but more expensive) version can get 22%. In the relatively near future, it might even be possible to get up to about 40%. The sun heats the land and the sea, causing air to expand and flow and causing water to evaporate, eventually leading to rain and snow. In these ways a small proportion of the sun's energy gets transformed into wind and water energy. We, in turn, capture a small proportion of that using wind and hydro turbines.

Why do renewables look even smaller on my graph than they do on some other graphs?

My renewables and nuclear figures are smaller than you sometimes see, simply because I have not multiplied them by a factor of 2.6. Why would anyone do that in the first place? The idea is that it takes 2.6 kWh of fossil fuel to make 1 kWh of electricity, because all power stations are inefficient. For this reason, applying the mark-up factor would be reasonable if all fossil fuels were used to make electricity. However, it isn't. Although all the renewables in today's energy mix can all be thought of as replacing fossil fuel in power stations, as renewables become a bigger part of our supply, this will cease to be the case. We will need to start using solar and wind energy for heat. For this purpose, a kWh of electricity provides no more heat than a kWh of fossil fuel. If I had applied the mark-up, as for example BP does in its annual review of world energy, you would see nuclear at 5%, hydro at 6.5% and other renewables looking slightly less pitiful at around 3%.[5]

Having seen what we use and where we get it from it is time for a quick look at the biggest problem with today's energy supply . . .

How bad are fossil fuels?

We owe a huge debt of gratitude to coal, oil and gas, but now we urgently need to stop taking it out of the ground.

Fossil fuel has powered us towards many of the good things in modern society, perhaps most importantly better health and life expectancy. But it is now clear that the carbon emissions from burning the stuff are a big problem and **we need to ditch fossil fuel just as fast as we can.**

Super high-speed summary of the evidence against fossil fuels

Extensive and detailed research by a very large number of scientists tells us uncontestably that emissions from the burning of fossil fuels is the biggest cause of human-induced climate change, and that this is on track to be very dangerous indeed for humans sometime this century – and no one knows exactly when or how bad it might be. Since it takes a long time to slow down and then reverse the causes of climate change, action is needed decades in advance of severe symptoms. Despite many years of debate on the need to do this, right now we are still *accelerating* the rate at which we pump greenhouse gases into the atmosphere. The hard-won 2015 Paris Agreement on climate change saw just about every country in the world (including the pre-Trump USA) agree that urgent action was needed. This simple and important milestone was a very long time coming, but still falls far short of ensuring that the required action will take place. We need to be phasing out the burning of fossil fuels at high speed and replacing it with clean or 'renewable' resources.

See the Appendix, pages 229–241, for the 14 things everyone needs to know about the climate emergency.

In the Appendix I have gone into more detail by outlining the 14 things I think everyone needs to know about the climate emergency. I have kept the detail out of the main flow because here I want to get straight into the follow-up questions. I want to go *wider* to look at things other than climate change, *deeper* into an even bigger issue of which the climate emergency is just a symptom and more *practically* to look at how humanity can find its way through some unchartered waters, and what any of *us* might do to help. That said, it might still be worth flicking to the back and checking you are up to speed on all the points listed, because they are essential, fascinating and not widely enough known. I have tried to put together a pithy summary. If you have a choice, please don't vote for any politician that hasn't got their head around all 14 points on my list.

Now we have seen that the fossil fuel has to go, we had better have a look at the alternatives. Some turn out to be amazing, some are limited and some useless or worse.

How much energy comes from the sun?

At any given time, there is a massive 16,300 kW of solar energy arriving on the Earth's surface for every person in the world.[6] That's enough for each of us to bring an Olympic swimming pool to the boil every day.[7]

Most of this energy lands on the sea, where it is largely out of the reach of humans. Under a third strikes the Earth's land area. That is still around 2,000 times more than we use and enough for each one of us on the planet to permanently have 2,700 kettles on the go all the time, or around 10 times the energy required to have the global population permanently in flight.[8] Plenty of energy, you would think, not just for us but for an abundance of other plant and animal life.

Can the sun's energy be harnessed?

Solar panels covering less than 0.1% of the total land surface (an area of 228 miles by 228 miles) could meet today's energy needs.[9]

Although currently well under 1% of human energy supply, photovoltaics are well worth getting excited about. This technology is coming of age much faster than many predicted, even just a few years ago. The growth rate in solar power has averaged a massive 50% per year over that past decade.[10] It has the potential to more than meet today's energy needs, allowing us to leave the fossil fuel in the ground, if only we would choose to do so. My land area sums were based on cheap solar panels with just 16%

efficiency although higher quality panels reach about 22%, and further improvements, perhaps up to 40%, could be possible within a few decades.[11] My sums also left aside, for now, a rack of technical difficulties that accompany solar power, of which energy storage, long-distance transmission and how to replace aviation fuel are perhaps the toughest. We will look at these challenges and will find that they are all surmountable. Nor are there any foreseeable shortages of materials needed to create the panels, nor any show-stopping environmental problems in their construction. While solar is better in some parts of the world than others, the overall abundance of the sun's energy means that most places should have enough for now. The availability of other sustainable energy sources such as wind and hydro are a further help for many of the places, like the UK and the Netherlands, that have high populations and not enough sunshine (more on this later).

If we could only keep the rate of installation up we could meet all today's energy needs in just 30 years. The price of solar power has been coming down too, by about 20% every time the rate of installation doubles, and this may well continue for a long time.[12] The problem is that as the scale goes up, it becomes a lot harder to maintain the growth rate. Already it is starting to fall to around 30% per year. That is still huge, but the declining trend is not good. Sceptics of the solar silver bullet point out that energy transitions of the past – from animals to coal, from steam to internal combustion engines and so on – have taken a long time.[13] The counter argument is that never before has there been such a globally acknowledged imperative to make a fast energy switch for the sake of humanity, not to mention many other species. That has to be able to make a difference, right? Surely, we humans are capable of some deliberate influence!

How much solar power could we *ever* have?

If 2.4% energy growth continues, in 300 years we will need solar panels to cover every inch of land mass. This would

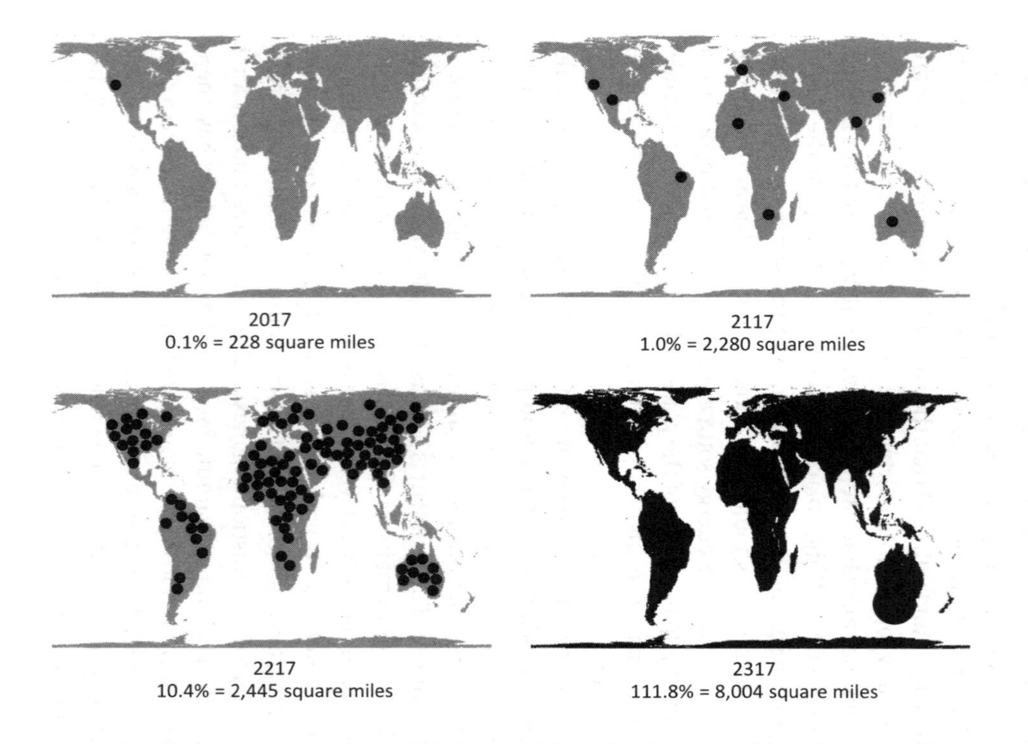

Figure 3.4 The proportion of the world that would need to be covered in solar panels if we stay on today's energy growth trend. The black dots are to scale.

leave no spare land for plants or animals: no sunbathing and lab food only.

Everything depends on whether, for the first time in history, we can succeed in deliberately limiting our energy growth. More energy means more solar panels. If the long-term trend of increasing our energy use by about a factor of 10 every 100 years continues, it will mean that by 2117 we would need roughly 1% of our land area. By 2217 we will need 10% and in 300 years we will need solar panels with today's efficiency covering just about every scrap of dry land, leaving none at all for growing food, and no opportunities for sunlight to be allowed to fall on human skin nor be used by other land-based species. All these sums ignore the usual spaces between the panels that we see on solar farms today. In short, either we curb our energy growth or we get used to a world that looks more and more like one huge solar panel. (And in my view, while we need them, they are about as beautiful as a car park.)

In theory it might be possible to increase the efficiency of solar panels by a factor of about three, which would buy us about another 50 years, ignoring again the need to use land for food. And if we could extend solar panels onto the entire surface of the sea, that might buy us yet another 50 years, taking us to roughly 2400.

If we can curtail the growth in human energy demand, then energy shortage can truly become a thing of the past without the need for fossil fuel, nuclear fission or fusion, or even unacceptable land use. If we can't control energy growth, then we will hit the limits and it is just a question of when.

Which countries have the most sunlight?

The top five are Australia, Russia, China, Brazil and the USA, and between them they hoover up 36% of all the sun that hits the world's land area.

In the map, the size of countries is proportional to the total sunlight, while the darker the shade, the more sun per person.[14]

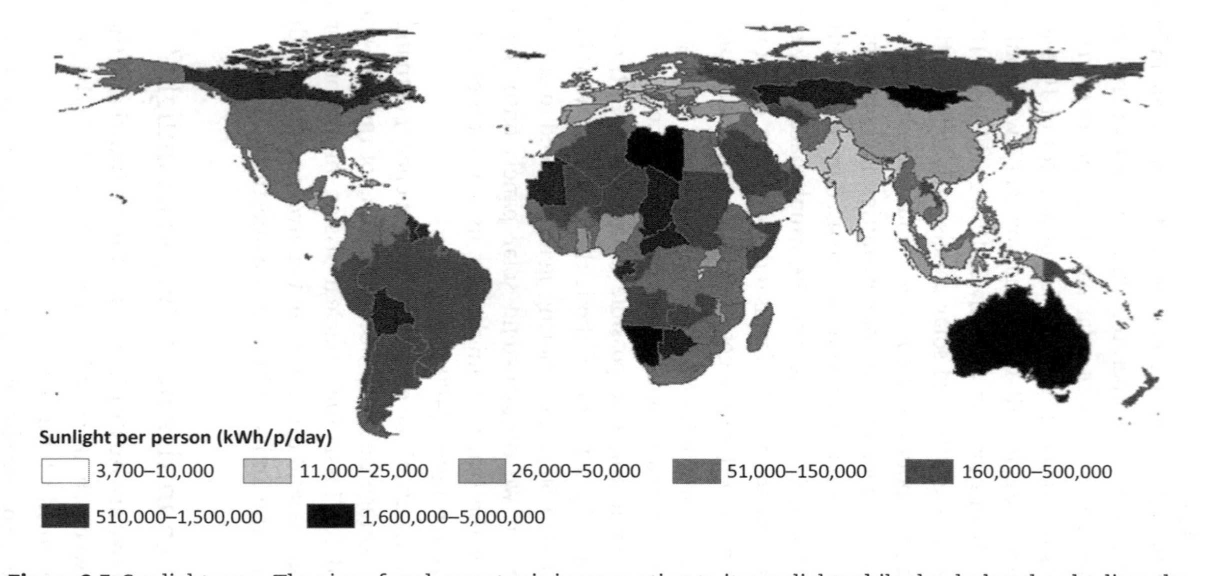

Figure 3.5 Sunlight map. The size of each country is in proportion to its sunlight while the darker the shading, the more sunlight per person.

It is a tight contest between the top five. Australia wins by having a lot of desert, coming in first with 7.5% of all the sun while the USA is in fifth place with just over 6%. Russia makes it into the group, despite being so far north, by being vast; 50% bigger than China and twice the size of the USA.

Total sunlight is a useful measure because it can give us ideas about who could be the world's big producers of renewable energy and food (which we look at in more detail later). There are complications though. Brazil had better not use most of its land for either solar power or food, because we need it to keep its rainforest. Much of Russia's sun falls on freezing wastelands that are hopeless for agriculture and require a lot of solar panels because the sun is so low in the sky. Australia's desert sun looks perfect for photovoltaics, but not currently for agriculture.

Sunlight *per person* gives us a feel, not of a countries' significance on the global energy and food stage, but of the relative abundance of energy that a country might feel in a low carbon world. Australia does well on this measure as well, with a whopping 200 times more sunlight per person than the UK. Put like this it looks strongly in the strategic interests of even the most neoliberal Australian government to push hard for a low carbon world. What they might lose by leaving their coal in the ground will surely be outweighed by their clean energy abundance.

On the map Africa looks unsurprisingly huge. In the main, it also looks encouragingly dark, although population growth looks set to turn it a good deal paler.

Which countries have the least sun per person?

The biggest losers in the sunlight per person stakes are Bangladesh, the Netherlands, South Korea, Belgium, the UK, Rwanda and Japan.

I have excluded from this list three countries with less than 10 million people. Poor Bangladesh not only potentially gets flooded as a consequence of climate change but receives only

3,700 kWh per person per day of sunlight. Even high-quality solar panels are only currently capable of capturing around 20% of the sun that lands on them, so if 8% of the land area was covered with solar panels they might just about achieve today's global average energy demand. My sums also assume no spaces between the panels, but there are more problems. Population growth will increase the burden, and even 8% of land lost to agriculture is pretty serious for a country with so many mouths to feed. It is hard to see how Bangladesh can develop without importing both energy and, as we will see later, food – both of which require exports of something else if the country hopes to make ends meet.

The UK,* crowded and fairly cold, also has just two and a half times as much sunlight per head as Bangladesh and its population has affluent, high-energy lifestyles. It has the advantages of greater potential from wind, wave, tide and even a bit of hydro up its sleeve. It has nuclear too (and that debate comes in a few pages' time). The UK's transition to low carbon energy looks doable, but compared to many countries, its solution will have to involve an interesting and complex mix.

On our sunlight map, India is looking pale because it has so many people, although sun per person is three times higher than in Bangladesh and 10% better than it is in Japan. Europe as a whole looks both pale and skinny.

What about when the sun isn't shining?

There are four basic solutions:

(1) Store energy from when the sun *was* shining.

(2) Supply energy from other sources.

(3) Make the demand match the sunlight.

(4) Transmit power around the world, because the sun is always shining somewhere.

* For the record, the UK is still a great place to live, despite the weather, the number of people and, in my view, a misguided sense of what democracy looks like.

The intermittency of sunlight creates challenges to manage, but not problems to panic over.

For short-term storage, battery technology and capacity is developing fast, and with it the idea of using batteries in both cars and households to deal with the problem of energy demand being out of sync with energy generation. Batteries can be made to fit into almost any machine, although their weight is a show-stopping problem for long-haul boats and planes, and their manufacture requires considerable resource extraction.[15] And even the best batteries wear out over time. Hydro storage makes an important contribution, although the potential for more is limited. Spare electricity is used to pump water uphill and then the pumps are run in reverse as turbines to generate electricity at the time when it is required. The process is efficient and, once built, a hydro scheme is a permanent solution that requires almost no material resources to run. But the scope of hydro is severely limited by the capacity of lakes, and attempts to extend this with new dams generally come with a huge environmental burden.

It is possible to convert electrical solar energy into the highly convenient and permanent form of liquid hydrocarbons, ready for use in most of today's (polluting) cars and planes. This is about 60% efficiency. That is a big loss, but it is not neccessarilly a show-stopper, unless those hydrocarbons then have to be converted back to electricity again, incurring yet more losses and taking the final energy supply to only about 20% of what we started with.

Hydrogen probably offers the most exciting potential for the future of storage to solve all the problems of intermittent renewable power generation. We can convert electrical energy into hydrogen fuel with 80% efficiency. Then it can be stored for as long as we like before using it to generate about 60% of the electricity that we started with. There is no great requirement for further extraction of materials. Hydrogen is also lightweight, opening up the potential at least for very long-distance transport, although its bulkiness presents problems.

A rack of other storage technologies are on their way, from fly wheels, to compressed air.[16]

In terms of generation when the sun isn't shining, wind, water and cautious biofuel all have supporting roles. For the medium term there might even be a limited case for a small splash of nuclear. See page 83.

In terms of demand matching to the sunlight hours, cars can charge during the day, large refrigeration operations can fit around the ups and down of the grid, and the timing of domestic use can often adapt to avoid peaks in demand. The smart technology to enable this and the pricing structures to encourage it are all achievable provided we get moving on them.

Finally, turning to the transmission challenge, China is investing in huge transmission lines to move electricity from one end of its country to the other. There are losses on the way but it is an increasingly doable exercise, even though transmission from places that are in daylight to those where it is night would mean even greater distances than across China.

Overall, the solutions we need to the problem of intermittency and storage are all coming along nicely. The critical factor is investment.

How useful is wind energy?

Of the sun's energy that lands on our planet, around 2% is converted into wind energy.[17]

And most of that wind is completely inaccessible, high up in the jet stream. On top of that, most of the low-altitude wind is far out at sea, even beyond the foreseeable reach of offshore wind farming. Even for the low-level wind on land and coastal waters, a host of geographical constraints ensure that only a small part can be harnessed. A wind farm larger than 10 km^2 can generate only around 1 W per square metre of land. That is down to the limit on the kinetic energy available in the air[18] and because, in a large wind farm, one turbine slows the wind down for the next. Even in a blustery place like the UK, while wind power is a help, it can't be the backbone of the solution. Turning the entire country into one huge wind farm would only generate 87 kWh per person per day.[19] Although that is nearly 50% more than today's global average energy use it is still less than current UK energy demand. By the way, in this unhappy scenario we are talking about turbines across Richmond Park, Scafell Pike, on top of St Pauls, . . . *everywhere*. The UK already has a passionate anti-wind lobby and even I would join it if things got this extreme.

Which countries have the most wind per person?

As we've seen, wind has nothing like the potential of solar power, but it is still capable of a useful contribution. The next map tries to give a hint of this, by looking at the kinetic energy in the wind above each country.[20] It is a very crude measure, and ignores the suitability of the terrain as well as the offshore options.

In many ways the sun and wind maps mirror each other with the population-dense sun-losers missing out on wind as well. Encouragingly, even without looking at offshore wind, of which it has plenty, Europe looks bigger and slightly darker when viewed in terms of wind rather than sun.

Overall, wind can become a significant part of the energy mix, although it is much more limited than solar power, and at least as temperamental. Some of the places with less sun can partially compensate by making better use of their wind and it helps a bit that the windiest days and months are often when there is least sunlight.

Why is sun better than rain?

Even if every raindrop was pushed through a turbine, capturing all the potential energy released on its journey from

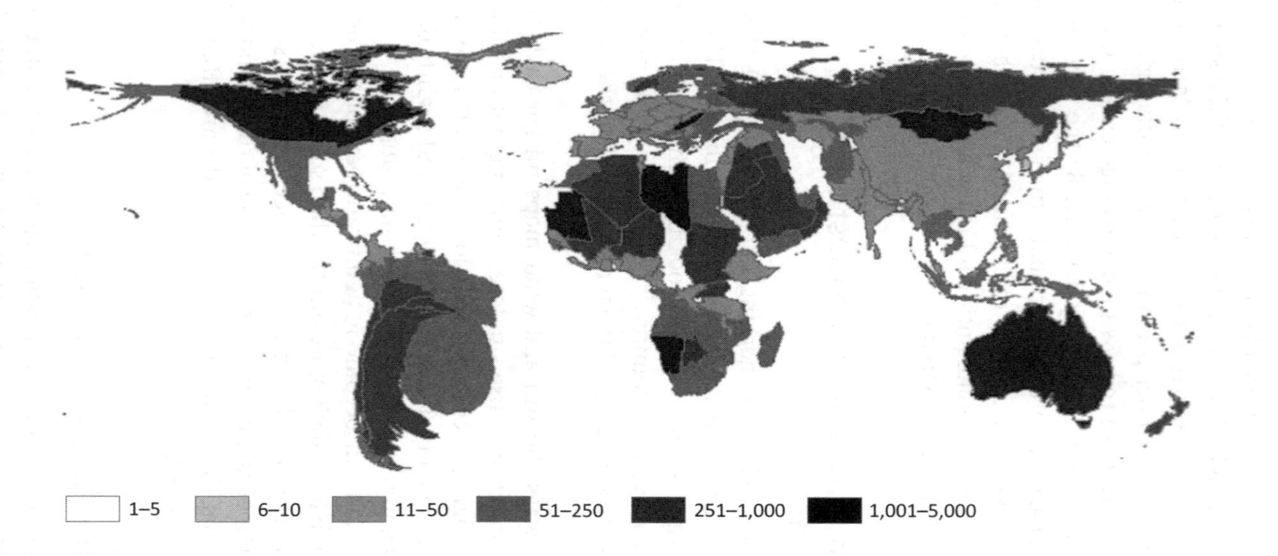

Figure 3.6 Wind map: the size of each country is in proportion to the total kinetic energy in the wind above each country while the darker the shading, the more wind energy per person. Offshore wind is not factored in.

where it landed to the sea, hydroelectric power would only just meet today's energy needs.

Global growth in hydroelectric power is unspectacular because it is starting to run up against the physical limits of geography. If you work out how much rain there is at what altitude, you can get a reasonable estimate for the theoretical maximum potential energy in all the world's rainfall. That is to say you can estimate the energy you might get if every raindrop went through a perfectly efficient turbine to capture all the energy available from its landing point on the ground down to sea level. It comes in at somewhere between two thirds and just over 100% of today's global energy usage, depending on whose estimate you use.[21] That makes it sound as though there may be plenty of scope for more hydro power until we remember that the idea of capturing all the potential energy from every raindrop is wildly unrealistic (thank goodness, as it would mean totally doing away with all mountain streams, and even, if you really think about it, all hillsides). Say, instead, we assume that 5% of that potential head of water could be put through turbines. That feels to me about as good as we could ever get, given the constraints of geography and remembering that big dams usually come with a serious environmental and social cost. And suppose those turbines were 80% efficient. On this basis, today's global hydroelectricity production, at 0.45 TW, is already around two thirds of the way to its maximum potential or perhaps more.[22]

My sums are rough but they are solid enough to show that whereas solar feels almost unlimited for this century at least, hydro is already approaching its maximum. We might see a modest rise but not a big leap forward.

Is nuclear nasty?

Yes. But the old arguments from the Cold War days need re-examining in the light of climate change risks and safety improvements. Trusted, impartial analysis is scarce but

essential to understand whether there is a role for nuclear in our medium-term energy mix.

A lot has changed since the polarised debates of the 1980s. On the positive side for advocates of nuclear power, our energy system needs intensive care like never before and the nuclear industry claims to have got a lot safer.

However, the half-life of nuclear waste is still just as long as it always has been. Once we produce it, it is with us for tens of thousands of years. It remains the case that a really bad accident or a deliberate piece of sabotage could haunt the whole world almost indefinitely. There have already been several near misses, and evolving terrorism capabilities to inflict such an event are an unknown. It is hardly an attractive proposition.

But another reality is that we are in a great deal of trouble with our energy supply. The fossil fuel really MUST stay in the ground. So, however nasty, let's take a look at some practicalities. Although nuclear looks wildly expensive compared to solar, a straightforward cost comparison misses the essential point that it offers a consistency of generation that neither solar nor wind can match. The UK and Japan are prime examples of countries where the sunlight per head is relatively low, and wind, hydro and wave are all chipping in to what could end up being a complex and erratic energy mix that might be made viable from a stable base of nuclear power alongside.

To weigh this up properly requires a highly complex analysis of costs, benefits, risks, opportunity costs and timescales. It needs to be made by people we can trust. They need to be able to assess nuclear's place in the wider context. We need to know that they have the right expertise, information and, critically, the right motivations to be able to make impartial and intelligent assessments. This is the crunch. The nuclear industry is not trusted, and there are historically good reasons for this. Across politicians and business, vested interests abound. In the UK, different bits of the same industry are currently not even collaborating properly between themselves, still less joining up with the rest of the energy scene to explore integrated solutions in the light of the full range of emerging technologies. The nuclear question is just one example of why there is a whole section on truth and trust later in this book.

Even setting aside the risks, we should keep our nuclear expectations limited. Over the next critical decades, nuclear would struggle to expand wildly from its current fringe position of supplying 1.6% of the world's energy. Currently there are more power stations reaching retirement age than there are new ones in the pipeline and building a new one is a far bigger deal than putting up a few million solar panels.

Anyone taking a firm anti-nuclear stance needs to have a coherent plan for the low carbon future without it. And equally, anyone advocating nuclear needs a coherent plan for how nuclear enables the carbon-free world and needs to show that this cannot be better achieved through the renewables and storage options that we have looked at.

Would fusion solve everything?

The answer depends on whether you would trust our species with unlimited energy.

One view is that fusion could be the great silver bullet to end humankind's energy problems forever. Bear in mind before sharing this view that it is the scale of our energy supply that has taken us into the Anthropocene, and all the dangers that come with it. More energy to solve the problem is like 'hair of the dog' – a beer for breakfast to fend off a hangover. I tried that only once, a long time ago, and don't recommend it.

Luckily, perhaps, we are still a few breakthroughs away from getting fusion off the ground as an energy supply. We don't know for sure how fast things might move forward but according to one group at MIT there could be fusion electricity feeding into the US grid by 2033.[23] One way of looking at it is that we have until then to get to grips with the Anthropocene.

Are biofuels bonkers?

The wheat required to power a Toyota Corolla on bioethanol for 1.1 miles could feed a person for a day.

That is a very steep trade-off between food and fuel. What I'm describing here is so-called 'first-generation' biofuel – in which edible crops are used to make liquid hydrocarbons. In the USA, far more food calories are used in this way than would be required to feed its entire population.[24] If we can crack how to make liquid fuel from cellulose, enabling us to use high-yielding energy crops (second-generation biofuels), the sums get better, but only by a factor of about five. I'm jumping ahead because in the transport section I will quantify the inefficiency of using land for biofuel compared to solar power. But it is important to note here that biofuels simply don't stack up as a large part of the transition from fossil fuels because for a pitiful return in hydrocarbons they place huge extra pressure on the already considerable challenge of feeding everyone. There is a role for high-yielding energy crops on some marginal land that can't be used for food agriculture. It is also possible to use crop residues. One major study of biofuel put the potential from these two routes as well as using waste food at what equates to around 20% of today's energy needs.[25] This sounds worth having until the side effects on biodiversity and soil quality are factored in. Use of marginal land requires great sensitivity not to threaten other ecosystem services and taking crop residues out of the agricultural system reduces organic content of soils.

Some have cited algae – supposedly the 'third generation' of biofuel – as the silver bullet. Micro-algaes grow incredibly fast, and are responsible for 40% of the world's current levels of carbon fixation. All algal species are able to produce oils. There could be millions of species of these microscopic organisms, potentially giving us humans a wealth of genetic resources to tweak and develop efficient biofuel resources. However, algae are far from commercialisation, with challenges ranging from species selection (breeding and genetics) to refining and processing of oils.[26] Exxon invested $100 million in this before pulling out. At best it looks too far off to put into energy transition plans.

So, to answer the question, biofuels in moderation are not entirely bonkers, but do need treating with a great deal of caution lest they become so. If carefully handled, they can provide a small but worth-having part of the energy mix. If an unregulated free market were allowed to run its course in the transition to a low carbon world they could be a disaster. We could see biofuel for the rich becoming more profitable than providing essential food for the poor, and we could see yet more natural habitats trashed in exchange for monocultures.

Finally, there is no problem with the idea of a car powered by used chip (french fry) fat except that the chip consumption required for this to be a significant help would have us dying in our billions from poor diets.

Should we frack?

Definitely not before we have a trustworthy analysis of the pros and cons and extremely good regulation. Only then might it *possibly*, *in theory*, be *marginally* worth doing. But almost certainly not.

Natural gas generates more heat for less greenhouse gas emissions than coal or oil. This opens up the theoretical possibility of using fracked gas as a transition fuel on the way to the low carbon world. The idea is that we use it to wean ourselves off coal and oil while the renewables get up to speed. Here are the BUTs:

- *First BUT* The gas we are talking about is methane, which, over a 100-year period is 25 times more potent a greenhouse gas than the carbon dioxide that we are trying to avoid. So if virtually any of the methane leaks out into the atmosphere, all of the climate benefits are lost. Serious leakage could make fracked gas far worse than coal. And I am not just talking about leaks at the time of drilling, I am talking about any breaking down of the integrity of the closed down, capped off well at any time in the future – ever. (Methane, incidentally, has a half-life in the atmosphere of only about 12 years. So it does all its damage in the early part of that 100-year period. This means that if, as we sometimes should, you look at a shorter timeframe like 50 years, the methane is almost 50 times as potent as carbon dioxide.)

- ***Second BUT*** It takes a lot of energy to run the fracking process itself – and this makes the carbon benefits marginal even if nothing goes wrong.
- ***Third BUT*** In the UK it will probably take 10 years to get fracking into production even if Cuadrilla – the main company driving this – was given the go ahead right now. By this time the UK is supposed to have already weaned itself off all its coal. We could sell the fracked gas to other countries to help them quit their coal too, but that option is not put forward by the fracking proponents and therefore feels unlikely.
- ***Fourth BUT*** Fracking involves the use of chemicals that we would not want to have contaminating our water supplies. In the USA this may be less of an issue as the operations often take place away from residential populations. But in the UK (and other densely populated countries) it is a big deal.

In order for fracking to be a good idea, we would need all these arguments to be assessed in a way that can be trusted. In this way the fracking debate is just like the nuclear debate. We need an analysis by people with sufficient expertise, who have access to all the information required, who have the resources to do the job and, critically, who are sufficiently free from vested financial interests that their motivations to carry out impartial analysis can be trusted. In the absence of this fracking should not be entertained. In the UK (for example), we are a long way away from this, and there are no signs yet of any movement in the right direction. The fracking issue is just one

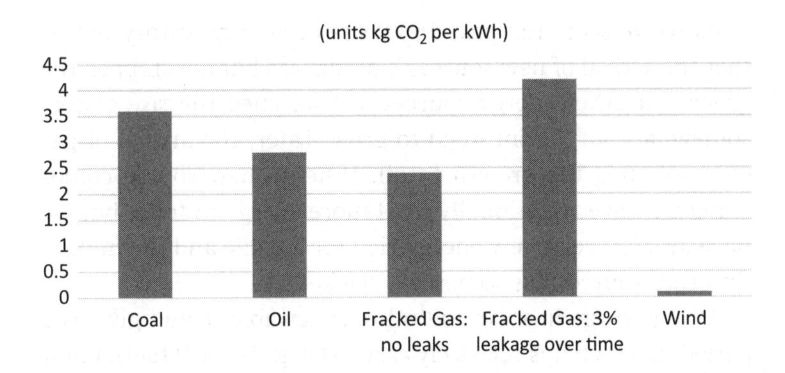

Figure 3.7 The carbon intensity of fracked gas relative to the alternatives. In the very best scenario fracked gas beats oil, but if just 3% leaks into the atmosphere over time, fracked gas becomes a lot worse than coal.

more example of the need for much higher standards of trust if we are going to find our way through the complexities of the Anthropocene. A further constraint, also requiring trust, is the need for superb regulations to be in place to ensure there are no leaks of chemicals or methane.

To sum up, given that even in the best case, the benefits of fracking could be no more than marginal in terms of greenhouse gas emissions, and given the trust-chasm that has opened up, the fracking answer is best left as a straight 'no', and attention turned to more fruitful areas.

Having looked at most of the options for a clean energy supply it is time to turn to the core dynamics of global energy use. This stuff challenges much of the accepted wisdom on climate action. It is totally essential to understand and yet is often and sadly either not grasped or is glossed over by energy policy makers.

Does more renewables mean less fossil fuel?

Not necessarily. The big question is whether we will have the renewables *as well as* or *instead of* the coal, oil and gas.

As we've seen, the past 150 years of energy history tell us that the arrival of new sources have dented but not stopped the growth of other energy sources. Oil softened the rise of coal somewhat, but it continued to grow. Later, the arrival of gas only softened the growth in oil. When a new source comes along we have traditionally used more energy in total, but we have also felt relatively energy-rich for a while and the hunger for other sources has somewhat slackened.

A huge surge in solar and other renewables could give us a period in which it is relatively easy to let go of fossil fuels, but it won't be enough to make it happen automatically. Policy makers need to get their heads around this. Please don't vote for any who haven't.

What is the catch with energy efficiency?

It goes hand in hand with an even greater increase in demand for whatever the energy is used for.

In 1865 William Stanley Jevons spotted that if the UK used coal more efficiently it would end up wanting more of it, not less.[27] This phenomenon has become known as the Jevons Paradox. Energy efficiency leads, by default, to an increase in total demand, rather than the decrease that is often assumed. It applies just as widely today as it did in 1865 and it has game-changing implications for energy and climate policy. It may be counter-intuitive at first but makes perfect sense on reflection. Look at it this way. Imagine if it took a tonne of coal to keep a family warm for one night and that family saves up to enjoy one warm winter evening – a New Year celebration perhaps. Now imagine that a more efficient burner is invented, and the same tonne of coal can keep them warm for two nights. Coal has just become twice as valuable to them, so they make extra effort to buy enough to keep themselves warm for three nights in the year. They might spend one of those nights fitting new insulation so that the coal becomes even more useful to them and the other night working by the fire to earn the extra money they need for their increased coal budget. However, the price of coal per tonne comes down a bit to help them because demand is going up so much and economies of scale are kicking in along with a stack of investment in new extraction technologies. And so it goes on. This is just a caricature of how the Jevons Paradox works, but I hope it demonstrates the principle.

Over the years we have become many times more efficient in our production of just about everything. LED lighting is hundreds of times more energy efficient than oil and gas lamps. Microchips are millions of times more efficient at storing data than paper and the cloud more efficient still. Electric trains are many times more efficient than steam trains, let alone horses. Yet our energy usage has risen hand in hand with those efficiencies and is actually *enabled* by them.

In fact, we can see that we don't use more energy *despite* the efficiency gains, but rather we are able to use more energy *because* of the efficiency gains. Wow! Feel free to pause at this point and reflect on the gigantic policy implications of this

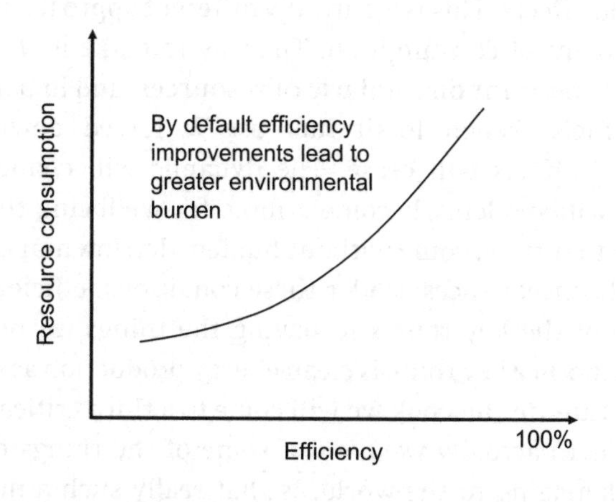

Figure 3.8 The Jevons Paradox.

perspective. It means that while efficiency gains help us get more benefit from any given amount of energy, they also end up leading to an increase in total consumption unless that is deliberately constrained.

Just before you go ripping out all your double glazing and deflating your tyres, note that I am not saying that efficiency gains cannot be useful in the future. But I *am* saying they are no good at all on their own.

(It is only fair to write that the Jevons Paradox has been hotly debated over the years. More detail on this and why the deniers are wrong is in this endnote.[28])

Given the catch, what can efficiency do for us?

We badly need more efficiency, but we also need to learn not to squander it with increased consumption.

We have to make efficiency work for us in a different way than we are used to. From now on when we get an efficiency improvement we have to deliberately bank the savings rather than allowing the default outcome in which our consumption appetite increases and the savings are lost through a myriad of rebound effects. This is a critically different approach to adopt at the point of consumption. The way to make it work is to have a **constraint on total use of resources**, and in particular fossil fuels. When fossil fuel use is forced downwards, rebound effects will cease. The dynamic will change. Efficiency will suddenly become a force for wellbeing that will, for the first time, come without hidden, detrimental environmental consequences. Under these conditions, efficiency will be one of the key routes to having the things we need and want. Another key route is clean energy production as covered above. Later in the book we will come to a third critical route, which is to actually *want less* of some of the things that are most damaging to the world. Is that really such a mad concept? We will see.

Is the digital economy enabling a low carbon world?

A popular storyline in the ICT industry is that the efficiency gains that it makes possible throughout the world more than compensate for its own carbon footprint – and that therefore ICT brings about the low carbon world.[29] It is true that digital information storage is millions of times more efficient than paper storage. And video conferences are thousands of times more efficient than flights for face-to-face meetings as a result. However, it is also true that because of the increased efficiency of information storage, we now store millions and millions of times more information – as well as keeping some of our paper storage. And while a video conference sometimes saves a flight it also sometimes marks the start of a relationship that *leads* to a flight that would otherwise not have happened. In the absence of a global carbon cap, an infinite number of rebound pathways eliminate any carbon benefits from efficiency savings. ICT enables more efficient logistics – leading, through rebound, to an even greater rise in transportation.

So the ICT industry's claim that it brings about efficiency improvements is true. This does makes a carbon cap easier to achieve without sacrificing quality of life. However, unless that industry pushes hard for that cap, it cannot claim to be enabling the low carbon world – quite the opposite in fact.

Now we've seen enough about energy dynamics to be able to look at what it might actually take to cut global emissions.

Why is cleaning our electricity just the easy part of the transition from fossil fuels?

Electricity from renewables can replace two and a half times the same energy in coal or oil going into a power station. Replacing heat sources is harder.

Different kinds of energy are useful for different things, but transferring from one type to another almost always incurs losses. In particular, to get electricity from oil or coal entails putting it through steam turbines at a power station where over 60% of the energy dissipates as heat that is usually not used for anything and less than 40% is turned into electricity.[30] Solar, wind and hydro power don't have this problem because they are in the form of electricity from the start. This means that if it is electricity that you need, a kilowatt hour of any of these is worth about two and a half kilowatt hours of coal or oil. This mark-up factor gives renewables a huge boost in the early stages of the clean energy transition.

A similar mark-up also applies to land vehicles for almost the same reason. The efficiency of electric motors compared to internal combustion engines means that a unit of electricity can power a car two or three times further than the same energy in the form of liquid hydrocarbon.

However, once we get to the point at which all our electricity is already coming from renewable sources, and all our transport has been electrified, we start having to use renewables to replace fossil fuel as a source of heat. At this point, they suddenly lose their mark-up advantage because a unit of electricity can heat a home or a blast furnace no more than the same energy in the form of coal, oil or gas. At this point the transition gets a lot tougher. When we hear anyone talking about the percentage contribution of renewable energy sources, we need to be very clear about whether it is the percentage in the whole energy mix or whether they are looking at just the easy bit; the electricity mix.

The final crunch in the clean energy transition comes when we start having to use renewable energy for things, like aviation, which currently require liquid hydrocarbons. If we have to produce liquid hydrocarbon from renewables we probably have to apply a mark-*down* factor of about 0.6. In other words, it takes about 2 kWh of solar electricity to produce just 1 kWh of aviation fuel.

How much fossil fuel can 1 kWh of renewable electricity replace?

2.6 kWh of coal going into a power station
1.7 kWh of gas going into a power station
2.5 kWh of gasoline or diesel for road transport
1 kWh of gas or oil for domestic or industrial heating
0.6 kWh of aviation fuel

How can we keep the fuel in the ground?

Since green energy on its own won't help much and efficiency on its own won't help at all, there is no escaping the need for a constraint on extraction.

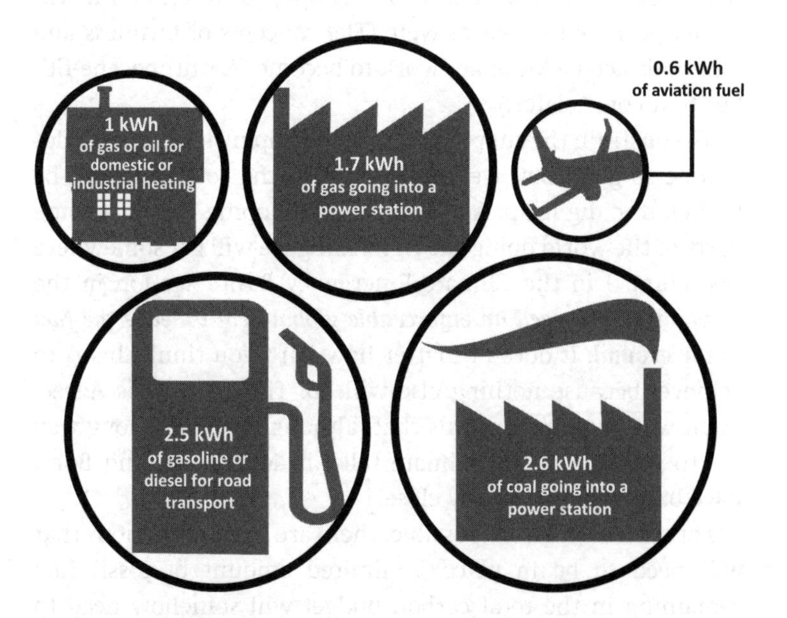

Figure 3.9 The amount of fossil fuel that 1 kWh (kilowatt hour) of renewable electricity can replace depends greatly on what we want to use it for. (The circles are to scale by area.)

The fossil fuel we use will be the gap between the clean energy supply and the total energy use. Straight away that gives us two clear levers; push the green supply up and hold the demand down. A third lever is to constrain the fossil fuel supply. This is the hard cap that I'm talking about, that will end the rebound effect on carbon emissions.

To push the clean energy supply up, we need to invest in it hard. This includes the rolling out of the renewable (especially solar) supply and the infrastructure to go with it, as well as the research and development into the rack of accompanying technologies that will be required to make solar work for us; storage, electric transport and so on. All this is doable. Where does the money come from? As luck would have it a whole lot of investment opportunity is created by the divestment from fossil fuels. The switch from yesterday's energy system to tomorrow's is loaded with business opportunities, and it will be net positive for jobs as well. (The concepts of business and jobs both need a bit of a rework to become 'Anthropocene-fit', but that comes later.)

To constrain the supply, it is no good hoping that renewables will be so good that we lose interest in the coal and can't be bothered to dig it up. And it is no good hoping that just some parts of the world doing the right thing we will get somewhere (see point 9 in the Climate Emergency Basics section in the Appendix). ***We need an enforceable global deal to leave the fuel in the ground.*** It doesn't matter how hard you think this is to achieve, because nothing else will do. The 2015 Paris Agreement was progress towards that, although leaving a long way yet to go. Subsequent climate talks in Marrakesh and Bonn have barely inched us any closer.

For such a deal to be possible, there are some conditions that will need to be in place. A limited amount of fossil fuel remaining in the total carbon budget will somehow need to be shared out and the very different ways in which countries will be affected by such a deal need to be taken into account, because unless it works for everyone, it won't happen.

The early stages of climatic change also look very different in different parts of the world. While the Maldives sinks, Bangladesh floods and California burns, Russia is likely to find its crop yields going up at first, its ports becoming ice free for 12 months of the year instead of eight, and yet more of its fossil energy reserves becoming accessible. In poorer countries a carbon constraint will certainly impact on wellbeing unless a clean energy replacement can be found, while in richer places the link between energy and happiness has probably already long been broken in the same way that the link between wellbeing and GDP has been shown to break down.

A global deal is going to be very hard to cut because it requires both understanding of the different implications for each country and a sense of international fair play that the world has never yet known.

The difficulties do not change the reality that we must have the global deal to leave the world's fuel in the ground.

The winners will have to compensate the losers. For this to be possible there will have to be a good enough understanding of the implications of a low carbon world for all parties AND there will need to be a strong enough sense of fair play and goodwill that such a deal can become possible. These feel like far-off conditions right now, because we are still locked in mindsets and habits that make the changes required seem impossible. But the later sections of the book look at how they might be achieved.

Other conditions that need to be in place are cultural and political and to do with the way we run society at the local, national and global scales. We come on to all this essential terrain in a few chapters' time.

Who has the most fossil fuel and how will they cope?

The five biggest players all have plenty of sun to use instead. Australia will be laughing, but Venezuela and a

few others are being asked to take a hit – unless this can be adjusted for.

The chart shows the 11 countries with the biggest fossil fuel reserves. Between them they hold over 80% of the world's proven reserves. My chart compares their share of the world's fossil fuels with their share of the sun. With caution, we can use it as one indicator of what the clean energy transition might mean for each of them as players on the world energy stage. It should give us some simple insight into the hopes and fears that the thought of such a transition might evoke. The top five, the USA, China, Russia, Australia and India all have plenty of sun as well as fossil fuel, so the transition might be as much of an opportunity as a threat. Australia in particular, as I've mentioned, will be basking in solar energy in exchange for its grotty coal reserves. Things also look good for Canada and Saudi Arabia.

On the other hand, Venezuela is being asked to swap a globally important position in the oil rankings for a relatively impoverished place in the sunlight rankings. It is fairly similar for Iraq and even Germany, which for a while led the solar revolution. This chart makes the low carbon world look extremely threatening to Qatar, which has plenty of oil but is too small to have much sunlight.

What does this mean for a global deal?

My comparison of reserves with total sunlight is by no means a perfect measure of the winners and losers. There are many other factors to take into account: the resources to make the switch, the availability of other non-solar energy sources, the 'usability' of their sunlight and so on. Sunlight on the equator is relatively compact and also comes in all year round, while in Russia and Canada it is more thinly dispersed and centred on the summer months. However, my charts do serve to highlight yet again that moving to a low carbon world means a completely different thing to different countries: some should be rubbing their hands together in glee while others are understandably scared. How could a global arrangement be made

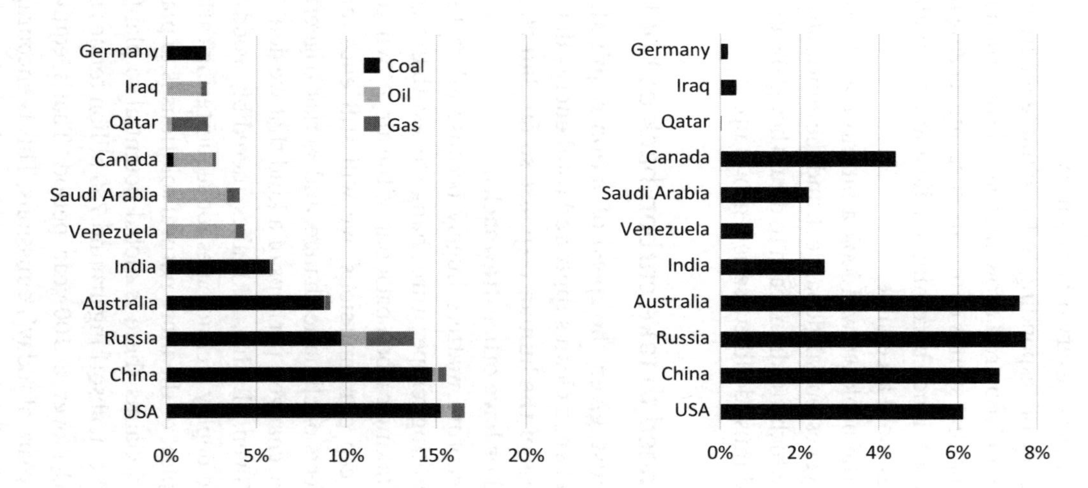

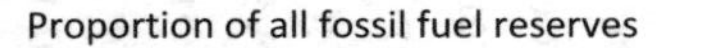

Figure 3.10 The countries with the biggest fossil fuel reserves, and their proportion of the world's sunlight.

without taking all of these different circumstances into account? Are we asking Qatar to be catapulted into poverty? If not, then a deal surely needs to see the world through its eyes along with those of every other country.

The urgency of responding to the climate emergency looks very different if you are a drowning island state than if you are a huge nation that stands to find its ports unfreezing in winter, its frozen wastelands becoming fertile and its abundant fossil fuel reserves more accessible.

Later in the book we will look a bit more at the seemingly impossible yet essential global deal, and the enormous question of ***sharing***, which children learn to do as they grow up and now humankind must do likewise as *it* grows up.

Will we need to take carbon back out of the air?

This is a must, given the speed of action required, the current gap between clean supply and total energy demand and the dopiness of the human response so far. But the options are limited or have other drawbacks.

Carbon capture matters hugely because all realistic scenarios for limiting temperature change to less than two degrees depend on taking carbon back out of the air. At whatever level we finally cap our emissions, we will still encounter some adverse effects of climatic change, and so risk triggering more catastrophic changes, perhaps of a kind that we don't even yet envisage. Given all this, it makes incredibly good sense to develop and deploy technologies for sequestering carbon.

The simplest of the more natural solutions is to **plant trees**. One study[32] claims there is global potential to plant about a trillion of them, absorbing around 750 billion tonnes of carbon dioxide (CO_2) over a 100-year period. That's equivalent to around 20 years of today's emissions. This is encouraging, but remember that it is largely a one-off win. Once the forests have matured, the rate of capture becomes much less. So if we play this card without cutting our emissions, we will be in trouble.

Other natural removal schemes include **soil carbon sequestration** through farming practices and **ocean planting** schemes, such as the replanting of seagrass in areas where the coverage has plummeted in recent years. All these schemes are healthy and important, but just like tree planting, they can only do a finite amount for us.

Carbon capture and storage (CCS) at the point of combustion doesn't really count as removal, but the technology is just about ready to go whenever the funding arrives. It is useful while fossil fuel persists in the energy mix but since it only deals with big sources of combustion, it will only ever be capable of capturing a modest proportion of our emissions. A step on from this however is **bioenergy with carbon capture and storage** (BECCS), the idea that you grow a biofuel and burn it, capturing the emissions and locking them away permanently. The concept was developed as an emergency measure to have up our sleeve in the event of catastrophic climate impacts but it has somehow found its way into mainstream climate mitigation pathways. We haven't demonstrated it at scale yet and it comes with the risk that the methods for locking the carbon away turn out not to be as permanent as we'd hoped – which could be disastrous. And, just like all the natural carbon sink methods, it turns out to have a finite limit, this time determined by storage opportunities. BECCS also puts pressure on the rest of our food and land system, thereby making the challenges discussed in Chapter 1 harder to deal with.

Direct air capture and carbon storage (DACCS) by mechanical means has the potential, in theory at least, to be a game changer, provided the additional renewable energy sources can be found to power the process. Sadly, we haven't yet managed to do DACCS at scale and there are still technological uncertainties. We don't know for sure whether an injection of long-overdue serious funding would lead to us being able to scale this up as fast as we learned how to make nuclear bombs or as slowly as we are learning to cure cancer. DACCS also comes with the same risk that BECCS has, that the carbon storage method may turn out not to be as permanent as we'd thought – which could be a disaster. In 2013, Duncan Clark and I wrote in *The Burning Question* that it would be unwise to rely on such an uncertain emerging technology. Now I write that, however uncomfortable this might be, we need to pull every lever we can and DACCS has to be one of them.

Biochar is the spreading of charcoal on fields to capture carbon. The storage here isn't permanent, but it might last for a few hundred years – which is probably good enough. It is another option with finite capability, but it could have the added advantage of enhancing soil fertility and thus helping with all the food and land challenges that we looked at in the first chapter.

Finally, **enhanced rock weathering** is another useful but limited option. Over a billion tonnes of CO_2 is already sequestered naturally every year through rock weathering. In the 'enhanced' process, finely crushed basalt or dunite rock is thinly spread on the ground, where it absorbs CO_2 at a faster rate. One study estimates the cost of this at just $60 per tonne if dunite is used, with potential for absorbing a massive 95 billion tonnes of CO_2 per year. That is more than twice current global emissions. The huge downside is that dunite has traces of harmful minerals such as chromium and nickel. Alternatively, at $200 per tonne of CO_2, basalt can be used and this has the added benefit of improving soil fertility by adding potassium.[31] However, the study estimates that the capture potential using basalt is much reduced at just under 5 billion tonnes CO_2 per year – well worth having but not a game changer. Rock weathering avoids the risks associated with deep storage of CO_2 underground or at sea, as well as the potential land-use downsides of mass sequestration through monoculture forestation.

We probably need a bit of all these methods and more, but to be very clear, no form of negative emissions is in any sense a substitute for action to avoid putting carbon up there in the first place.

Can carbon be offset?

Since the scope for negative emissions is limited, there is no substitute for cutting emissions in the first place. Any 'net-zero' targets need to be split into two parts – the emissions-reduction bit and the carbon-removal bit. Otherwise one gets traded against the other with emissions allowed to remain, and consciences are salved with a spurious 'offset'. Most offset propositions don't really stack up, but there are one or two at the very expensive end of the spectrum that just might.

The idea of a carbon offset is that you pay money to fund a carbon-reducing activity to undo the damage of your emissions. The 'voluntary offset' market is the one that we hear most about. The idea is that you can call your flight, your lifestyle or your company net-zero carbon because you pay money into a scheme that somehow cuts carbon elsewhere by the same amount as your own footprint.

The recent fashion (which I am not knocking) is of declaring a climate emergency and setting net-zero emissions targets. Many people and organisations are asking whether the easiest way of avoiding too much real change to lifestyles and business models might just be to 'offset' their carbon rather than cutting it in the first place. It is a tempting concept that is sometimes dangled in front of us by airlines and others at mouth-wateringly affordable prices. In 2019, the global voluntary offset market surged to around $300 million, claiming to offset a whopping 100 million tonnes of CO_2.[33] That's just $3 per tonne of carbon saved. At prices like that, $6 could make good the climate damage from a return flight from London to San Francisco and the average UK person could offset their entire lifestyle for just $40 (£30) per year. If it were real, I'd be delighted to sign up right now and have done with all my carbon concerns. And the whole world could do the same for a trivial $170 billion per year; 0.2% of world GDP. It feels too good to be true, and sure enough, on closer inspection, it turns out to be dreamland.

All the so-called 'offset' schemes need to be assessed against a few simple tests:

(1) Do they *actually* remove carbon from the air? (Sounds too obvious to have to say it – but there are plenty that don't.)
(2) Can they be *verified*?
(3) Are they *additional* to what needs to happen anyway? (And this test is very hard or impossible to pass since nearly all the negative emission methods turn out to have finite potential and we are in such a mess that we already need them all.)
(4) Do they come with show-stopping *environmental costs or risks*?

Let's have a look at some of the most popular types of offset schemes, starting with the cheapest but sadly most bogus, and working upwards to the most robust but sadly also most expensive.

Some of the cheapest schemes don't take any carbon out of the air at all but rely on very dubious claims to help someone else cut their carbon. Examples include buying more fuel-efficient stoves for African villages, protecting forests from deforestation that should not take place anyway, and funding renewable electricity that needs to happen even without you or your company incurring its footprint.

Next cheapest and often far more credible are schemes that involve plants, and especially trees, removing CO_2 through photosynthesis. Yet verification that a forestation scheme has actually taken place can be a nightmare, especially if your project is on the other side of the world. Even with a rack of certification schemes to choose from, it is hard to know whether your money has really gone where you think and whether you are the only one claiming the offset for a given tree. Supposing these barriers can be overcome, planting trees is undoubtedly important. But as we have seen, another problem with such measures is that they are finite, so if we use them as a substitute for cutting emissions in the first place we will be left in a high carbon world that has burned up all its negative emissions options. For example, in January 2020, Sky (now part of Comcast) went for the underwater equivalent of forestation, announcing seagrass planting as part of a net-zero strategy, giving them what they think is a bargain-priced offset option. This too has finite potential.

In 2020, the brewery and restaurant chain Brewdog announced that alongside a strong decarbonisation programme for both its operations and its supply chains, it would, with immediate effect, take twice the remaining carbon back out of the air through carefully vetted and ecologically sensitive forestation schemes, including one on its own land in Scotland. For me this is a leading example of a company taking action on climate. I'd better declare an interest because I'm very happy to have advised on this and my staff screened 65 certified forestation projects before finding a small handful that we were happy with.

After the natural removal techniques come the more technical solutions for air capture. These come with varying blends of high costs, risks and finite limits.

If carbon storage really can be made permanent and risk-free, and supposing there really is no immediate limit to the world's renewable energy capacity, then direct air capture has the best chance of passing all the tests above and therefore being the most defendable carbon offset. The problem is that rather than costing $3 per tonne, the average price in the offset market, DACCS comes in at more like £1,000 per tonne and, as I write this, is only commercially available through one small pilot scheme in Iceland, which sequesters a tiny 50 tonnes per year.[34] That said, the good news is that various pioneering companies are developing the technologies and, if you believe them, the price might foreseeably come down to under $100 per tonne. Moreover, in theory, there is less of a limit on the amount of carbon we might be able to remove from the atmosphere through DACCS.

Summarising how to think about offsets and net-zero targets:

(1) Cutting carbon is *essential* and negative emissions are not a substitute. All the so-called offsets mentioned here can be good things to fund – but this is a separate activity to carbon cutting and it is probably misleading to use the term 'offsets'.
(2) Any net-zero targets need splitting out into a carbon-cutting target and a carbon-removal target. Otherwise there is a strong temptation to use offsets as an excuse to shirk responsibility for mitigation.
(3) All carbon-removal schemes have a finite limit, with the possible exception of DACCS – which is risky, very expensive, still experimental and increases the global energy requirement.
(4) Despite the risks, DACCS is probably the most legitimate 'offset' mechanism, for anyone determined to use that word. Natural removal methods, if done well and carefully verified, may be acceptable in the short term.

How much energy are we on track to use in 2100?

A continuation of 2.4% growth per year would mean we will be using seven times as much in 2100 as we are today.

Given the history, a continuation of the age-old pattern of growth has to be the most likely outcome unless something very fundamental happens to change the underlying dynamics. More specifically, 2.4% looks as likely a rate as any. You could argue for lower, based on the very long-term average rate being lower and perhaps reading some significance into the growth rate having dipped a tad in the past decade, or you could argue for higher, based on the growth rate having climbed steadily over the past few centuries.

Two things might change that dynamic. One is a major crash of human civilisation due to our failure to deal with our arrival in the Anthropocene. This book is about that not happening.

The second is that our species matures. We come to understand that 'growth' is no longer about getting *more* in terms of physical power. We learn to be more careful, to get our kicks without smashing the place up and perhaps even to be a gentler species. A new, wiser humanity such as this would only grow its energy supply and usage if it were confident it could do so without adverse consequences.

I don't have a crystal ball, but I'd be surprised if one or the other of these scenarios didn't happen this century.

Many institutions create elaborate energy scenarios and sometimes even forecasts. Some show energy use topping out of its own accord.[35] Some people have suggested to me that human appetite for energy might just naturally run out of steam without us having to be deliberate about it. Their theory is that we might simply be reaching a point at which, without any great change of mindset, we find we already have all the energy we could wish for. Advocates of this view point to trends in some developed countries which show energy growth subduing or consumption even declining and suggest that when the poorer countries catch up, everyone might have had enough. I'm afraid that I think all this shows a staggering lack of imagination of the ways we might find of wanting more energy. I expect there were also some ancient Egyptians who thought that if only they had 100 times as many slaves their energy needs would also have been met for all time. To give just one example of how innovation and efficiency might lead to us wanting to consume energy this century, the possibility of desalinating water for agricultural use looks like it is being made more viable by graphene filtration technology. This opens up a huge potential for farming in the desert. It's brilliant but comes with a colossal new energy demand. A different kind of example can be seen in the early forays into space tourism.

The final flaw in the 'natural peak' argument is that it is simply not possible to infer global energy trends by looking at only certain countries, since the dynamics of energy growth play out at a global scale, and whatever goes on in one place effects and is affected by what happens elsewhere.

Can enough energy ever be enough?

For the long term, this is the huge question. Somehow we need to become content not to grow our energy supply. Individuals and communities have achieved it. The challenge is to do so at the global level.

Some people I meet say energy has always gone up so it always must. Some say it will come to a natural halt on its own. Some say our species will wipe itself out because we are so unfit for the situation we are in. Some say we will be fine because we always have been. (By the way, the dinosaurs could also have said that before they died off.) Some wishfully tell me that solar power will replace fossil fuels because the market will simply make it happen. Some say the transition to clean energy has to take a long time because all transitions always have done. Some say global cooperation is beyond the capacity of humankind. Some say levels of inequality are determined by 'human nature'. Some say humans wake up only when they have a nasty experience, so we may as well wait until the inevitable nasty experience happens to us (and they are generally talking about exceptionally nasty experiences).

All this strikes me as determinism. I am not a determinist. I reject any scientist telling me they can prove the absence of free will. I don't actually think anyone can prove it either way, and I don't see much point in a long discussion about it either. Like most people, I have at some point contemplated the possibility that I'm a figment of my own imagination and, of course, I can't rule it out. However, my practical observation is that people who live as if they have free will tend to have more positive lives. To put it another way, determinism is a way of dodging the challenge of trying to do things that are difficult.

It requires effort to change long-engrained habits. And the dynamics of human energy growth are very deeply engrained.

Humanity is going to have to raise its game if it wants to take deliberate control over the amount of energy it uses.

As this book goes on we will get deeper into the underlying issues that might enable the change that the evidence tells us we need to see. We will explore whether we are capable of such change. I think there is probably everything to play for. It certainly isn't proven that we can't make the shift. But we will have to learn to think in new ways. We have to get into this terrain, however tricky it is, because at some point our big species is going to have to grow up.

Energy solution summary

- We need to ***both*** leave the fossil fuel in the ground ***and*** grow our clean energy supply to replace it. The latter will not cause the former.
- The **world needs a deal** to leave the fuel in the ground. The Paris Agreement is a statement of intention but is a long way short of a firm arrangement to make it happen.
- Efficiency improvements are very important but on their own they do not generally bring about reductions in energy requirement. In fact the opposite effect is more usual; efficiency gains tend to be accompanied by an even greater increase in output, so that the total energy demand goes up not down.
- There are no show-stopping technical barriers to the clean energy revolution.
    - **Solar** is the world's best renewable opportunity by far, and has enormous potential, despite its current tiny contribution to the energy mix. It will need to be rolled

out much faster than any energy transition the world has ever seen.

  * Supporting technologies are also needed but they too can arrive in time if we invest properly.
  * **Wind and hydro** are useful but limited.
  * **Nuclear is risky**, permanently polluting, expensive and very hard to roll out quickly, but given the mess we are in, it may still have a place in the energy mix, especially in countries that have less sunlight per person and because of its reliability compared to some of the other sources, and because we may need nearly every source we can get.
  * Biofuels on any scale require extreme caution as they threaten both food supply and biodiversity.

  While those technical solutions are only broadly sketched out in this book, I hope it is enough to make it clear that they exist or are emerging fast – and no net suffering or hardship is required. In fact there are opportunities to live better as we make the transition.

- Given the scale and urgency of the challenge, and the pondering nature of human response so far, we will almost certainly need to **take carbon back out of the air**, even though these technologies are not yet fully developed or understood.
- The more we can **limit our energy demand**, the easier the transition to clean supplies will be. At some point we will have to end energy growth. We need to learn that enough can be enough.

Energy: What can I do?

I'll keep this to a few simple things, as you have probably already had a lot of advice on personal energy management. More importantly, think of it as a big systemic challenge. Personal action is all about influencing the whole system.

- Vote for politicians who both get the issues and prioritise them. If none do, vote for the ones who come closest. Let the candidates know what you are looking for.
- Where you can, spend your money in support of energy-efficient supply chains, low carbon technologies and infrastructure. Examples – if you need to buy a car make it electric if you can, insulate, put up solar panels, push for your pension to divest from fossil fuel and invest in the solutions we need.
- Get better at enjoying things that don't require much energy. Examples include walks, books, most types of party, most forms of socialising, local holidays and any hobby that doesn't rely on fossil fuel.
- Decrease energy consumption in all the ways that you already know about or can easily get advice on elsewhere. Examples include energy-efficient home life, moving from big gas-guzzling personal cars to small, electric and/or shared transport, flying less. Think about the embodied energy in the stuff you buy. Don't buy junk and make the good stuff last.
- Without losing friends, or beating anyone up, be challenging of habits and views that are unhelpful, at work and play.
- Don't beat yourself up. Have fun while you cut your energy footprint. Find ways that make life better. Pick your battles. Accept that you will always be flawed and keep going.

This is a short list, and the idea is that you do them all.

If only energy and climate were simply technical issues. In order to get the global deal that we need, we are going to need many other things to be in place. This takes us into issues of global governance, fairness, economics, growth, inequality, truth, trust, values and even how we think. We will turn to those things later, but first we need to cover one more group of physical issues, travel and transport. This is inescapably intertwined with food, land, sea, energy and climate.

4 TRAVEL AND TRANSPORT

Travel and transport accounts for a big enough chunk of the world's energy use with most of this coming from liquid fossil fuels. These have the huge convenience of being mobile and dense energy stores – ideal for cars, boats and planes. So replacing this will present a raft of technical and infrastructure challenges. All these will be made easier the less travel we actually need.

We start with an overview of world travel before getting into just a few of the specifics.

How much do we travel today?

The average person travels 3,921 miles per year: 57% of these are by road, 23% on foot, 7% by rail and 13% by plane.

The walking miles include everything from people hiking the Andes to you walking between the fridge, the kettle and the cupboard as you make your morning cup of tea.[1] Since the turn of the century all these modes have been growing. Walking has been going up by just 1.3% per year, roughly in step with population. Air miles have been rising by a massive 7% per year (doubling every 10 years) and car miles have been going up by 4.7%.[2]

Cycling, by the way, which we often think of as an important mode, turns out to be much less significant to human mobility than cars, planes, trains or feet. It makes sense when you consider that many people in the world don't cycle at all, most of those who do only use a bike for trips of up to a few miles and yet almost nobody rides a bike between their fridge and their kettle, or any of the other tiny journeys that make up so many

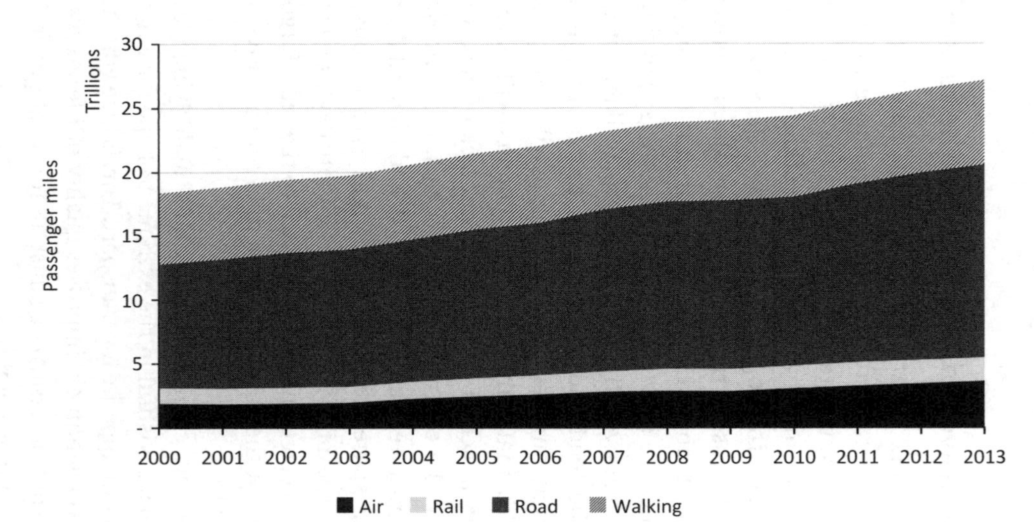

Figure 4.1 Annual total global travel by mode of transport.

of our walking miles. The average Dutch person cycles 1.4 miles per day,[3] but in Spain, where 73% of people say they never get on a bike at all,[4] the average is just 100 metres per day. In the UK it is slightly higher at 265 metres per day.[5]

How much travel will we want in the future?

Some people think we are getting towards an end point in human desire to travel further. In rich countries, they say, some people are getting to the point where they don't want to travel any further, and that when the rest of the world catches up, and provided the population stabilises, there will be no need for any increase in transport. I can see the attractiveness of the argument, and I'd love to believe it. It is certainly true that aeroplanes are inherently grotty places to spend time no matter what class you fly, what films are available or whatever is on the food and drink menu. But overall, as with arguments that our demand for energy will also plateau, I fear the idea that our travel requirements will naturally top out shows a lack of imagination and lack of observation of evidence to the contrary. As I write this, the Chinese are building a high-speed rail link in the UK and Richard Branson is developing space tourism. Meanwhile autonomous cars open up the possibility of endless extra journeys, delivering everything you can think of, right down to the kids' forgotten gym kit, without the natural constraint of human travel time being required.

Unless we can change humankind's underlying growth dynamic, it looks wishful in the extreme to hope for a natural trailing off of our appetite for transport. This is true whether or not more travel genuinely improves quality of life. As we will see later, I'm not necessarily against all forms of growth, but peak-travel would be helpful.

Once we stop using fossil fuel, most transport will require land, whether that is to generate electricity from solar panels, to grow crops for biofuel or simply to grow food to power human legs.

The relative sustainability of different transport options will have a lot to do with the number of miles per square metre of land that each option might deliver. Here's how some of the most obvious possibilities stack up:

How many travel miles can we get from a square metre of land?

A square metre of photovoltaic (PV) panels in California could power an electric car for 1,081 miles per year or an electric bike for a massive 21,243 miles.

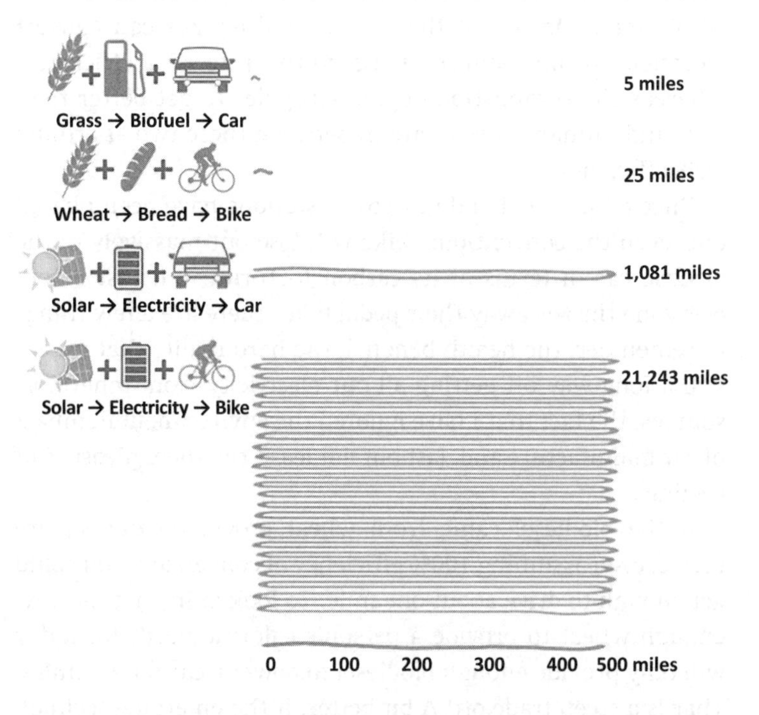

Figure 4.2 Distances are based on using a small car powered by biofuel (global average yield); cycling powered by eating wheat (again, based on the global average); and driving a Nissan Leaf electric car or riding an electric bike, both powered by Californian sunlight. (For the electric bike I've assumed no human pedaling at all.)

Alternatively, we could eat the wheat grown on our square metre and use it to walk 13 miles per year or cycle for 25 miles.

At best, a car powered by biodiesel from willow grass could only get about five measly miles.

Overall, the electric bike and electric car numbers here are enormously encouraging. Their incredible efficiency compared to conventional cycling and biodiesel starts to make intuitive sense when you consider that even cheap solar panels can convert about 16% of the sun's energy into electricity whereas photosynthesis in plants is generally not capable of capturing more than 2%, and even less gets converted into edible food. On top of that, e-cars and e-bikes can convert electricity to mechanical energy with up to 80% efficiency, whereas the combustion engine struggles to get better than 25% and human muscles are in between these two at around 60% efficiency.

Once all our coal and gas power stations have been phased out, even the conventional bike will lose out massively to the electric car in terms of its carbon footprint. But just before everyone throws away their pedal bikes there are a few things to remember; the health benefits, the hard reality that we are still a long way off getting all our electricity from renewable sources, the fact that I have ignored the environmental impact of car manufacture and, last but not least, the sheer pleasure of cycling.

If the biodiesel came from wheat grown on our square metre, even assuming 100% efficiency of conversion, you could get enough to drive about one mile. To look at in another way, enough wheat to provide a person's calorific needs for a day will only provide enough biodiesel to power a car for 2.7 miles. That is a steep trade-off! A bit better, if the emerging technology can be cracked, would be to create biodiesel from willow grass or some other cellulose crop. Even if this could be made to work it would still mean only around 20 car miles in exchange

Table 4.1 Travel miles possible using energy from a square metre of Californian land that is used for either solar panels or growing wheat or willow grass.

	Miles per square metre per year
Electric bike (PV)	21,243
Electric train (passenger miles, PV)	4,033
Nissan Leaf electric car (PV)	1,081
Tesla electric car (PV)	927
Bike (bread powered)	45
Walking (bread powered)	22
Airbus A380 (passenger miles) – (biofuel from cellulose)	12
Biodiesel train (passenger miles, wheat powered)	5
Biofuel car (willow powered)	5
Horse-riding (wheat-fed horse)	3
Biofuel car (wheat powered)	1

for a day's food. A very important message for all energy policy makers to grasp is this:

> ***Whatever its low carbon credentials, biodiesel should not be thought of as a major part of our energy solution. Going down this route would put a huge strain on the global food system and risk widespread malnutrition.*** ***(See Are biofuels bonkers? on page 88.)***

More details of the calculations and a list of caveats are in the endnotes as usual.[6]

How can we sort out urban transport?

The ideal city is compact and easy to get around. The buildings are tall and generally close together, except for the recreational green spaces. Going out from the centre, houses are not too big and definitely not detached (since there is no need for this from a sound point of view and it just wastes space and makes them harder to keep warm). In this way most able people can get where they want to on foot or by bike most of the time.

Sadly, most cities have already been built and retrofitting this is not easy. Silicon Valley is a needless sprawl of low-level detached properties, many with multiple garages, and the shops each have their own car parks. The result is that where kids could be walking between each other's houses and everyone could be no more than five minutes' walk from the countryside, hardly anyone goes anywhere except by car. Having made such a mess of the town planning, it is hard to re-engineer things. San Francisco, just nearby, has been saved by the space constraint of sea on three sides and is alive with pedestrians and cyclists, despite its ferocious hills.

Most cities largely have to live with the building stock they have got and make the most of it whether it is good or awful. That leaves three strands of action for sorting out the transport. First up is to clean the vehicle fleets. Bikes, e-bikes, e-cars, and e-everything else. Second is the transport infrastructure; bike routes, walk ways, trains above and below ground, buses, trams, and park and rides. Thirdly, the IT that joins up the travel modes needs to make it so that the lowest-energy solutions are as easy to arrange as reaching into your pocket for car keys. As well as the fixed route public transport, IT can also sort out the flexible vehicle sharing with taxis adjusting their journeys and prices in real time, negotiated with customer needs and adjusting prices accordingly. In this way the total number of vehicles on the roads can go right down, along with most people's requirement for ownership. It is all doable. In fact, in 2019 Oslo banned cars from its city centre, and Copenhagen and Amsterdam are already ruled by bikes.

Will shared transport make life better or worse?

Worse if you are committed to always driving your own vehicle and keeping it full of your own personal junk. Much

better if you want to be able to travel in the most appropriate vehicle more of the time, without incurring the overhead, and are open to making social connections with other humans.

As long as you haven't lost them down the back of the sofa, there is no denying the simplicity of reaching for your car keys to start your journey. For this reason, the private car has been the default choice for the disorganised. Your own personal car is also hard to beat if you like to have a lot of your own junk with you when you travel. And you never have to think about driving a car you are not used to. But the price we pay for our own dedicated capsule is that we are stuck with the same vehicle whatever journey we are trying to make (unless you own a few), we can only start each journey from wherever our own car is parked up, and we have none of the environmental or cost efficiencies of shared ownership.

In today's world, most of us own cars and hiring is usually a relatively awkward process. It takes time, isn't spontaneous and usually requires pick-up and drop-off from inconvenient locations. So, if you want to take the train for part of your journey, it is usually hard to have access to a car at the other end if you need one. The global car share market is still languishing around the $1 billion mark, making it a pin-prick on the whole car scene.

It doesn't need to be like this. In the car sharing world, the car share infrastructure is massively improved, and the public transport options are better too, and the apps to make it all fit together have come of age. If you need a vehicle, you can quickly access the ideal one for the journey you need, from a bike or e-bike to a small run-around that can always find a parking space right up to a van for moving furniture. You can always take the smallest car that meets your need for the trip and when you are not using it, it isn't your responsibility. Bliss. You can also take the train knowing that at the other end you can jump into a shared car if you need to. In this world, there are fewer cars but each one is put to better use. Of course, there are far more bikes. The air is cleaner. Northern European cities currently lead the way.

Should I buy an electric car?

Firstly, ask whether you still need to have a car of your own at all. If you don't, don't buy a new one until you need to. Then make it electric if you can and as economical as you can. Make it last.

Roughly two thirds of the carbon footprint of driving an oil-powered car is down to the fuel and the rest comes from the emissions involved in manufacturing the car in the first place. So, unless they are very inefficient, cars should be looked after and kept on the road for a long time.

Electric cars cause less emissions in use than their oil-powered equivalents even if the electricity has all come from a coal power station, because their engines are so much more efficient than internal combustion engines. This more than compensates for the inefficiencies of electricity production at the power station and the very high emissions from burning coal. However, the overall gains are marginal and apply only if you buy the most efficient car you can. Sadly, it doesn't work to think of your car running on renewable electricity even if you sign up to a deal that tells you this is true because, for the foreseeable future, there aren't enough renewables to go around. So, if you use them up, everyone else has to have more coal in their electricity mix.

If long journeys or lack of charge points makes the fully electric car impractical for you, you might find that you can get away with the kind of hybrid that can go 50 miles on its battery and only uses fuel to power a generator if needed. In practice 90+% of your miles will probably end up being electric, as long as you charge up when you can.

Cleaning up our transport isn't just about carbon and energy, but also about immediate human health.

How urgently should I ditch my diesel?

Urgently if you do a lot of urban miles.

A mile of congested urban diesel driving takes about 12 minutes of life away from the rest of the population – that's if you add up the all tiny impacts you have on every other person.

In the UK 40,000[7] people a year die prematurely from air pollution, and vehicles cause 8,900 of those deaths. That is five times more than the 1,775 who died in road traffic accidents.

This question is a bit of an aside but, as a cyclist, I can't resist it. The stats will make you want to hold your breath every time a bus goes past.

The two main killer pollutants are small particles and nitrogen dioxide (NO_2). Diesels belch out far more of both of these than either petrol or electric cars. If, like me, you cycle or even walk in cities sometimes, it is only natural to be curious about how these pollutants work.

In terms of particles, the smallest ones (often called PM2.5, meaning less than one 400th of a millimetre across) are the most harmful because they find it easiest to get into your blood stream when you breathe them in. Diesel cars typically produce about 15 times as much of these as their petrol equivalents.[8] Some particles (around 10% of those from vehicles) are also emitted from brake pads, tyres and road surfaces, so even electric cars contribute slightly to the problem. PM2.5 typically last in the atmosphere for days or weeks, depending on the weather. If it is windy they disperse faster but can also be blown in from elsewhere. Rain washes them away, while in dry weather traffic can pick them up off the ground and swirl them back into the air. People often think trees will help filter them out but sadly it is the reverse. Although leaves are surfaces for the particles to land on, a much more important effect is that on a busy high street trees form a wind barrier that holds the particles down at street level.

If you do a lot of urban driving, every particle coming out of your exhaust pipe stands a reasonable chance of going straight into the lung of a passer-by, whereas if you are on a country road, they are dispersed and are much less likely to encounter a human before settling out of the atmosphere. So, where you drive makes a big difference to the health impact.

Unlike PM2.5, nitrogen dioxide doesn't get formed instantly. When air gets very hot, such as in a car engine or even in a gas flame or wood fire, nitrogen oxides are created. These only turn into the harmful nitrogen dioxide some seconds or minutes later when they react with ozone or oxygen in the air.[9] This delay is good news for anyone walking around Times Square or Oxford Street or on a bike stuck behind a bus, because it means things get a chance to disperse a bit before getting to you. The bad news is that nitrogen dioxide tends to last longer than particle pollution and isn't washed away by rain, so while there is still more of it in cities than in the countryside, wherever you go the air we breathe contains nitrogen dioxide contributions from all over the world.

Electric cars have no exhaust pipes but, if the power comes from the grid, there will have been plenty of both particles and nitrogen dioxide coming out of the power stations that generated it.

I hope the figure below helps you to decide what you personally want to do. It shows the life-minutes taken away from the rest of the population per mile driven for different types of journey and different vehicles.[10] A five-mile drive in a congested city in a diesel car takes a person-hour away from the lives of the people you drive past.[11] So if you are a London taxi driver, things look relatively clear cut to me. Swapping your black cab for petrol, or better still an electric version, would do everyone else a favour. Ideally, get something half the size too.

Some diesels are a bit cleaner than others. I can't finish this section without mentioning the Volkswagen Emissions Scandal. MIT estimated that the insertion of emissions-cheating software by VW into its diesel cars – the ones sold in Germany alone – will result in 1,200 premature deaths over the lifetime

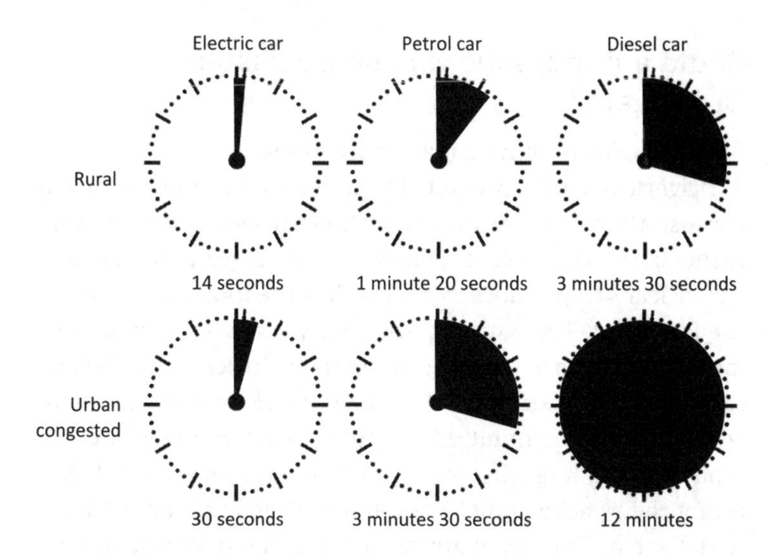

Figure 4.3 Life-minutes taken away from the rest of the human population per mile driven. I have assumed all vehicles are medium-sized (1 tonne) fairly efficient cars.

of the vehicles.[12] Worldwide, the figure must be much higher. If there is any difference between mass murder and this method of knowingly causing people to die, it seems to me to be a thin one. Some of those responsible have had their wrists slapped and lost jobs. Why we take this kind of crime so much less seriously than street stabbings or drug dealing seems odd to me.

Diesels became popular in the UK for their supposed carbon efficiencies. However, this benefit is marginal because while diesel cars do a few more miles to the gallon, a gallon of diesel also has about 20% more carbon in it.

Taking carbon, particles and nitrogen dioxide all into account, there is a clear hierarchy with diesels at the bottom, petrol in the mediocre middle and electric cars at the top by a long way. Whatever type you chose, it's good if you get a small one, drive it less and share it more.

Could autonomous cars be a disaster? Or brilliant?

It all depends on how much we use them.

Driverless cars undoubtedly stand to be more efficient because they can slip stream each other and optimise every manoeuvre. The first difficulty, as we've seen, is that efficiency leads to yet more trouble unless the total global carbon use is capped. The issue is particularly extreme with autonomous cars because they also stand to be far less stressful and safer. We could sleep on the way to work or sleep all night while it drives us hundreds of miles to a meeting. We could send our kids huge distances to school every day, and if they forgot their packed lunch we could just send the car out again to deliver it. The opportunities for increased energy use are enormous.

Before we slip into this way of living by default, we should ask ourselves very carefully the essential but slightly weird and unusual question of whether the quality of life will be made better or worse by this innovation. Just because we invented it, it doesn't mean we are forced to adopt it, even though as we will see later, it may be hard to resist the pressure. Is the experience of being alive in a driverless car going to be better than the experience of being behind the wheel, or not in a car at all? My instinct is that once the novelty has worn off, it will be almost as inherently dull as frequent flying.

The question of autonomous cars takes us back to two questions. Can we cap the carbon? And, more widely, can enough be enough? If the answers are yes, then autonomous cars can help us towards sustainable living in the Anthropocene. If not, they will only make things worse.

How can we fly in the low carbon world?

An A380 carrying 550 passengers from New York to Hong Kong burns through 192 tonnes of fuel. That is about 36% of

the weight on take-off. Without using fossil fuel, the challenge is to carry enough energy on board.

Low carbon flying is a hard nut to crack, but solutions look possible. The most obvious one is to use biofuel. Unfortunately, we saw on page 88 that the trade-off between fuel and food is incredibly steep. If the biofuel used was made from wheat, the amount of it required would be enough to meet the calorific needs of all the passengers for a whopping four years and would require 1.5 square miles of prime California land for a year to produce it.[13] To put it another way, the wheat required for meeting today's aviation needs equates to about 2,100 kcal per person per day for everyone in the world – almost as much as humanity's entire food calorie requirement. If we can crack the technology, it would be something like four times more efficient to grow a willow crop like miscanthus and make biofuel from that, but the numbers are still hideous enough to make flying very difficult.

Luckily, there may be a better way. The problem with electric planes has always been that batteries don't have the fabulous energy density of fossil fuel, so they make planes too heavy to fly. In fact, it takes about 20 kg of premium lithium ion battery to carry as much energy as there is in a single kilo of jet fuel. For this reason I had written off the idea of electric planes. Until, that is, I heard that Easyjet are planning just that for their short-haul flights, and I decided to give things a closer look. The A380 that we have already looked at is about the most efficient commercial plane the world has ever seen in terms of the energy required per mile per tonne of payload. And it has the advantage of being huge. To find the best solution I've played around with the natty flight simulation model that was built by my friend who is a physicist, pilot, software engineer and air traffic control advisor. First of all, I've stuffed the A380 full of batteries instead of fuel. Next, I've assumed that an electric plane can do a better job of converting stored energy into motion than a conventional jet plane (a factor of 2.5 feels reasonable, given the inherent inefficiency of wasted heat when burning fossil fuel and the efficiency improvements that electric cars achieve over oil-powered ones). The result is a plane that could happily manage the 600-mile flight from London to Berlin. We are getting there! Furthermore, future improvements in the energy density of batteries will result in a proportionate increase in range. For emergencies, a bit of spare energy is needed, and the best way to carry that might still be as fossil fuel, even if a generator is required to convert it to electricity should the need arise. Electric flights are not so barmy after all.

A third option, and perhaps this will be the best for long-haul flights in the foreseeable future if the solar revolution takes off as hoped, is the idea of using solar electricity to create aviation fuel from carbon dioxide in the air. There is about a 40% energy loss in the conversion process, but never mind. I estimate that to get our passengers from New York to Hong Kong would require about 25 m^2 of solar panels for a year per person, provided you put them in a suitable bit of desert.[14] That is 270 times more land-efficient than the biofuels-from-wheat option.[15] My numbers are only very rough, but good enough to deliver a clear message. What a great opportunity for Australia, which will never look back from ditching its coal industry.

So, it looks like flying can be possible in the low carbon world. Please don't confuse me saying this with the idea that we don't need to think about how much of it we do. Remember that the amount of fossil fuel we use will be equal to the gap between our total energy use and the amount of non-fossil energy we have, and planes use a lot of energy. So, please, if you have a choice, don't vote for any politician that talks about airport extensions without giving proper air time to the huge environmental impact of flying.

Finally, solar panels on the wings will always be strictly for adventure seekers rather than those trying to get from A to B.

Should I fly?

Whether for business, love, fun or gap year it all depends ...

There is no denying the massive carbon footprint. For perspective, London to Hong Kong and back economy class is about a quarter of the average UK person's annual carbon footprint.[16] That's not me trying to beat you up, it's just fact. Whether it is worth it depends on why you go and what you make of your trip. It's a trade-off.

If you fly on business, it might depend on whether or not your work is helping to build a sustainable future for the world. It might depend on whether you are going to come back with a wider view of the world and a greater sense of global empathy. If your nearest and dearest live on the other side of the world you might have a dilemma. If you just want a holiday, then it is between you and your conscience, but if you do decide to go, it is essential that you have a fantastic time.

Kevin Anderson is a professor of climate policy at the Tindal Centre, UK, and is someone whose work I greatly respect. He takes a strong no-flying stance. If he has a climate conference in China, he takes the train. It makes a point and is a shining example of authentic leadership. But on the other hand, to me, his work is good enough that any flight that helps him to do his job is probably worth it.

So my thoughts are that if you do fly (and I sometimes do), at least treat it as a very special occasion and extravagance. Fly economy (to take up less of the plane) and make the most of your trip. If you are travelling to see the world, make sure you go for a long time, have new experiences and meet people who see, think and live differently from those you have been surrounded by at home.

Do virtual meetings save energy and carbon?

At the moment, no. They are slightly more likely to trigger a flight than to save one. They make a low carbon world

more possible, but they do not, in themselves, help to bring it about.

I have done a lot of work for big tech companies that are tempted to argue that video conferencing technology reduces flights and thereby saves millions of tonnes of carbon throughout the world. But if you've read the book so far you will already have guessed that the rebound effect undoes this argument. To illustrate the point, here's how I came to make quite a few flights between England and California. Some people in Silicon Valley read one of my books and liked it. They made contact by email. We had a few phone calls and a video chat. We got on so well that they decided they'd like me to do some work for them and six trips across the Atlantic followed. (All justified in the name of trying to encourage a tech giant to do its bit to enable the conditions under which the world might agree to leave the fuel in the ground.) Without the virtual meetings, the real ones would probably never have taken place.

Once again we see that, at the macro system level, innovation, technology and efficiency gains are just part of the process by which total resource consumption goes up, not down.

However, to look at things the other way round, if the world were to adopt the global carbon constraint that it needs so badly, virtual meetings would be one of the critical technologies that would enable us to conduct our lives and businesses in the normal way.

How bad are boats? And can they be electrified?

Sea freight is over 30 times more energy efficient than air freight – but you need to be a lot more patient.

A cargo ship carrying a 15,000-tonne payload at 15 knots typically gets through 470 tonnes of fuel on the 6,000-mile journey from Hong Kong to London.[17] It therefore requires a kilowatt hour of energy about every one and a half metres of its journey.

Or, to put it another way, it requires about 0.07 kWh per tonne per mile, provided you keep the speed down. So about 20 apples, oranges or bananas can be brought to you by sea from that far away for about the same energy cost as driving an average conventional car for one mile. Sea freight, in fact, is an essential enabler of the whole global economy. It doesn't just allow dense populations in cold climates to eat food grown in sunny spacious places, it probably enabled the supply chains of most of the things you can see around you right now as you read this. Doubling the speed to 30 knots would cut the journey time from two weeks to one but since water resistance goes with the square of the speed, the energy use would go up four-fold. So, patience is good.

By contrast, the air freight alternative requires about 2 kWh per tonne per mile,[18] but can get there the same day. The tiny portion of goods that travel by air are mainly luxuries (such as asparagus in winter and fast-fashion clothing). Crisis management in poorly-managed supply chain logistics also resorts to air freight. (In a former life, working for a boot company, air freight was the very expensive emergency option when stock management had gone astray.)

However, while the energy and carbon footprints of shipping are good 'bang for buck', we've already seen that we need to cut out the fossil fuel altogether before long. And a conventional container ship uses the gunkiest, most polluting form of oil that there is; a black tar that can't be used in cars or lorries.

Here is how the basic physics of electric sea freight stacks up. As with planes, a much greater weight of batteries than fuel would be required for boats because the energy density of batteries is about 20 times lower. But the difference is that for boats, the weight of fuel is a much smaller proportion of the total weight. Even if the electric motors were no more efficient than conventional ships engines, it would be possible to make the journey with just over 9,000 tonnes of batteries on board, allowing 6,000 tonnes of cargo. That's a 60% reduction in payload but not enough to make it totally unfeasible. If, as is likely, an electric motor can be twice as efficient as a conventional ship's engine, then it would be possible to carry just 4,500 tonnes of battery and 10,500 tonnes of cargo. Before jumping to the conclusion that we can easily roll this out tomorrow, remember that it puts yet another strain on global battery supply.

Sails and solar panels on deck, by the way, are nice ideas and potentially worth having, but turn out to produce only marginal gains. The problem is that today's boats are huge compared to the wind and sun energy that they can capture on board.

Finally, a note on passenger travel. It is tempting to think that this could be the eco-friendly option, but sadly, much of the inherent efficiency of boats compared to aeroplanes is thrown overboard because people don't like to lie obligingly side by side for the journey, like apples and bananas do. The difference becomes extreme on a luxury cruise when everyone has to have not just a cabin, but access to swimming pools, casinos, dining halls and all the rest. (I've never been on one, but I've heard.) This takes the carbon cost to a massive 0.22 kg per passenger mile, on a par with driving a small petrol car as the sole occupant[19] or flying. Human sea travel still stacks up for relatively short sea journeys, sailing boats, pedalos and rafts. (See this endnote for three highly recommended reads.[20]) (See also page 34, Is local food best?)

E-bikes or pedals?

Both – thank goodness.

If I had to write that the most sustainable way to get around could be by electric bike rather than the old-fashioned ones with pedals, I would be doing so with great sadness, and a sense that technology was leading us into a worse world. A few pages back we learned that a piece of land with solar panels on it can power an electric bike about 200 times further than the same land used to grow food for a conventional cyclist. Sad but true.

Happily, there is room for both. E-bikes make practical a whole new range of low carbon journeys that are not possible on a conventional bike. They stand to relieve congestion, pollution and noise, and use radically less energy than all the four-wheel alternatives. They are, for example, a massive opportunity for smart city businesses. Having a fleet of these will help your staff turn up to meetings on time, uncrumpled, sweat free, and probably happier, while you save money on the side.

One note of caution, however. There will need to be a new level of attention turned to cycle safety. At the high-powered end of the scale, an e-bike could be as fast and dangerous as a motorbike, as well as harder for pedestrians to hear coming.

This is emphatically not the death-knell for the good old pedal bike, which keeps you fit and comes with a wonderful simplicity. Perhaps I'm unusual in that all my clients tolerate the sweat and creases that I turn up to my meeting with, and in exchange for that I like to think that they benefit from the increased vitality that I get from having had the exercise.

Finally, while looking at travel, there is one piece of dreaming that needs a reality check lest it gains any more traction.

When might we emigrate to another planet?

Humanity's entire energy supply would be only enough to send just one small manned spacecraft per year on an intergalactic journey.

It could be tempting to hope that if we could move to a Planet B, the need for 'one planet living' would only be temporary. It might mean that we could get away with hanging on to the same expansionist mindset that has allowed our predecessors to move into almost every corner of the Earth. Some people, including one or two eminent scientists, have said things that suggest it is just a question of time before our species finds its way onto other planets.[21] But basic physics quickly knocks the idea into touch.

The nearest planet that might have conditions reasonably similar to Earth is Proxima Centauri B. It is local in space terms – just four light years away. So if on-board entertainment were so good that we could think of 40 years as an acceptable journey time, we'd only need to travel at one tenth of the speed of light to get there. But even to go at that speed, a person needs a ginormous amount of kinetic energy. In fact, humankind's daily global energy supply is about the same as the kinetic energy of getting 50 people, weighing 70 kg each, travelling at this speed. And that is just them, flying through space in their underwear to Proxima Centauri B.[22] But since this would mean instant death, a spacecraft is required. To keep things as optimistic as possible, let's say we make it weigh just 600 kg per person. That is the same as my small Citroen C1 car, which I can safely say I would not want to be stuck in for 40 years. This is travelling super-light when you think that the Space Shuttle weighed 80 tonnes and that we haven't factored in any of the kit and food required for the journey or for getting started in the new land. We also haven't yet allowed for any energy losses in the propulsion mechanism. Finally, it would be wise to think about how we are going to slow down at the other end, so we don't hurtle into our new home at 30,000 km per second. That will almost certainly require energy, which will also need to be carried with us. Once all these considerations are taken into account it is already clear that even one small spaceship with a handful of people would require humanity's annual energy supply, leaving nothing for those left behind on Earth.

Clearly this whole discussion is fantasy, but my simple sums make the point that interstellar travel is not an alternative to 'one planet living'. Star Trek was a fiction programme. We cannot currently or for the foreseeable future, and probably not ever, travel at 'warp speed' or through a wormhole.

Much more possible, but a lot less useful, might be to migrate some of us into Earth's orbit or perhaps to Mars. This would mean living permanently in leak-proof containers. It would be a lot simpler, but just as nasty, to do this on Earth in a mine or a concrete bunker.

Much more possible still would be to start polluting the universe with capsules of our DNA, in the narcissistic hope that somehow they take root. Perhaps better beings than us will get hold of our genetic code and be able to recreate us (but whether they would bother is another question). None of this will help in any way with the practical challenge of life for us humans on Earth or for the next few generations.

Whatever we make it into, Earth will be our only home for a very long time to come. There is no Planet B.

5 GROWTH, MONEY AND METRICS

We've seen how it is physically possible to move quickly beyond fossil fuels and still have our energy needs met. We've seen how everyone can have a healthy diet while improving environmental dimensions of land and sea management. We've even seen how we can meet our transport needs at the same time. We've touched on biodiversity, disease risks and plastic pollution. All these challenges turn out to be happily solvable from a scientific and technical perspective.

If only that were enough.

This book, remember, is really about finding ways to make life much better than it is now. Planet-saving is just the first-aid part of that project. For both the realisation of better life and for crisis management it is clear that economics needs a major re-examination. Given the radically new situation that we find ourselves in, it is hardly surprising that some of this stuff which has been refined over millennia but for a totally different context might not still be fit for purpose in the Anthropocene. I'm not exactly the first to make this observation.[1]

Does our economics stem from our values or is it the other way around? A bit of both perhaps, but certainly the more our world is framed up around financial gain and individualism, the harder it is to think in more cooperative ways.

I am not an economist, and that makes it easier for me to see things afresh, but also means that I'd better not attempt too much detail.

We will explore some of the types of growth we have been experiencing and pursuing to ask which types are still healthy for us today. In looking at money we will explore how markets, investments, wealth distribution and the ways in which we spend can help or hinder us in addressing the challenges we've seen in our air, land and sea.

A lot of mainstream economics is going to be challenged.

Which kinds of growth can be healthy in the Anthropocene?

Children get physically bigger as they grow up. Adults, if they want to keep growing healthily, have to find non-physical forms of growth. Humanity needs to undergo a similar transition.

It is clear that a big dynamic has been going on over which humans have so far had little or no agency. Innovation brings new technologies and efficiencies, enabling greater energy use and yet more innovation. This has been the mechanism of growth and expansion for a very long time.

Until recently, humans have always been able to get away with a physical growth mindset: growth of energy, population, infrastructure, life expectancy and money has all been fine even if it has gone hand in hand with growth in carbon, mineral extraction, pollution and just about every human impact you can think of. But here in the Anthropocene it is suddenly no longer clear what things it will still be healthy to grow and what things will not. Rather than clutching on to growth as an essential or rejecting it as the root of all trouble, we need to dissect the growth question and ask ourselves, in today's world, what *kinds* of growth are desirable and which are not.

At one end of the spectrum between healthy and unhealthy growth there are some forms that have become as useful to us as cancer. At the other end are some human characteristics that we need to inflate like life jackets, as quickly as we can. In between are some chocolate teapots and some non-essentials. I've plotted them onto a chart with a suggested direction of transition. And I hope you won't mind me having manipulated the shape of the line to make it into a traditional exponential curve, purely as a comfort to any traditionally-minded economists who may be reassured by the concept of scampering up an exponential curve.

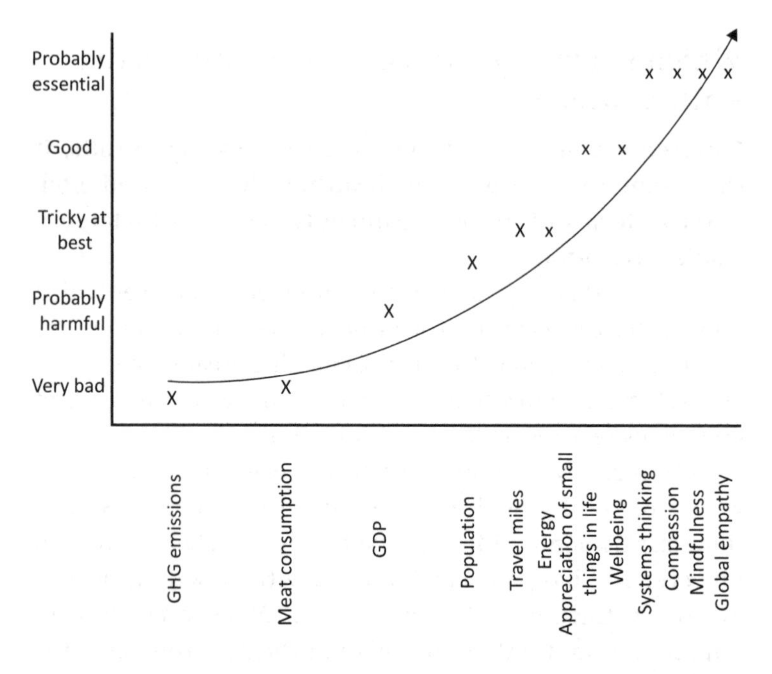

Figure 5.1 Imagining humankind's transition to maturity expressed as the changing nature of our growth. The exponential shape has been made up purely as a comfort for any traditionally-minded economists.

Here is my assessment of some of the various types of growth.

Greenhouse gas emissions: Harmful since they exacerbate the climate emergency.

Meat consumption: We've already seen the risk this poses, to humans and many other plant and animal species.

Energy use: Not inevitably harmful but very dangerous since it is linked to human capacity to trash the planet, including but not restricted to the climate emergency. We have seen that sometime in the next few hundred years, unless nuclear fusion comes of age, we will also run up against the finite limit of the amount of energy landing on Earth from the sun.

Consumption: Harmful. By this I mean consumption of just about everything physical.

GDP: Irrelevant at best. To date this has been linked to both carbon and energy growth and those who think the link is inevitable logically draw the conclusion that GDP growth must halt too. However, at heart GDP is just an abstract human construct that has no inevitable connection to any physical activities. It would be physically possible for everyone to go about their life in exactly the same way without the exchange or use of money at all.

On the other hand, I'm not sold on the idea that GDP growth is the root of all our problems either. It would be possible, in theory at least, to grow GDP simply by charging for new services that have lower physical impact (examples include neighbourly care, other forms of human contact, and cloud computing), while phasing out or charging more for higher-impact activities like petrol cars.[2]

Population: A bit more growth can be tolerated provided we distribute it properly. Since the impact that each person has depends so much on how they live, it is clear that 1 billion reckless people would trash the planet in no time while 15 billion very careful people could live in harmony with a thriving environment. The more people there are the more careful each one needs to be. We do not need to panic about the idea of 12 billion people, although fewer would make life easier. There is one proviso to the idea that 70% population growth is survivable: we will need to distribute the resources to the people. So, either the people are going to need to move to where there is sufficient food and energy production, or those resources are going to have to flow to them. Specifically, we have seen that existing agricultural production can feed the projected 9.7 billion people in 2050, but that only works if a lot of food grown in North and South America flows to Africa.

Flights: Harmful for the foreseeable future. We saw on pages 126–127 that the pressure that aviation puts on emissions is incredibly high, even if electrification could be made to work for both short- and long-haul flights. Every policy maker needs to understand this. Any who don't are surely unfit for office.

Technology: Good if, and only if, we become more selective in what we develop and how we use it. We have to end the default pattern of developing stuff because we can, and then someone using it whenever it gives an efficiency improvement, and then everyone else having to use it because others are doing so and it is uncompetitive not to. A continuation of this dynamic will be catastrophic in the future. We need to control our technology, rather than it controlling us! More on this in a few pages' time.

Life expectancy: Good – provided of course we are talking about *quality* life years. Our capacity for keeping people alive looks set to sky-rocket, so we are going to need to be more discerning about when those lives are still worth living. It seems to me that we are going to need new debates on how and when people are allowed to die and on the way in which the resources to keep people alive are prioritised. It is a brutal reality that every life saved by a health service carries an opportunity cost. We would do better to look at this clearly than pretend it isn't true. Worth remembering too that increased life expectancy carries with it an implication on population growth.

Wellbeing: An infinitely superior goal than GDP growth. Whether it lends itself to measurement is another question. In fact, the exercise of trying to strangle human wellbeing into quantifiable metrics could be self-defeating.

Awareness: Crucially important. Appreciation of small things could be the most critical and essential thing for humans to be growing. Awareness of every aspect of the human experience: it is not what we do or see or have that matters, but how much we appreciate it. (I'm not claiming to be especially great at this by the way.)

To sum up, we need a radical overhaul of human growth aspirations. We need to change the flavour and shape of our growth. We need to grow in our maturity, awareness and compassion. We need to grow in our capacity to appreciate what we have and what is around us. This is not in any way the end of ambition, but is a shift in its nature.

Some of these things we might grow deserve looking at in a bit more detail . . .

Why is GDP such an inadequate metric?

GDP goes up if things we used to give and receive for free become chargeable. A country's GDP might well go up if the level of spontaneous kindness were to go down. If friends won't baby sit or look after a frail neighbour, then it becomes a commercial activity. To give another example, the laundered profits from drugs and other crimes show up as GDP.

In 1968 Bobby Kennedy summed it up in a speech that is inspiring and clear enough to deserve a long quote:[3]

> *Too much and for too long, we seemed to have surrendered personal excellence and community values in the mere accumulation of material things. Our Gross National Product, now, is over $800 billion a year, but that Gross National Product – if we judge the United States of America by that – that Gross National Product counts air pollution and cigarette advertising, and ambulances to clear our highways of carnage. It counts special locks for our doors and the jails for the people who break them. It counts the destruction of the redwood and the loss of our natural wonder in chaotic sprawl. It counts napalm and counts nuclear warheads and armored cars for the police to fight the riots in our cities. It counts Whitman's rifle and Speck's knife, and the television programs which glorify violence in order to sell toys to our children.*

> *Yet the gross national product does not allow for the health of our children, the quality of their education or the joy of their play. It does not include the beauty of our poetry or the strength of our marriages, the intelligence of our public debate or the integrity of our public officials. It measures neither our wit nor our courage, neither our wisdom nor our learning, neither our compassion nor our devotion to our country,* ***it measures everything in short, except that which makes life worthwhile****. And it can tell us everything about America except why we are proud that we are Americans.*

How do our metrics need to change?

'What gets measured gets done' said the 1980s management guru Tom Peters,[4] and his words have been repeated so much that they might easily be misunderstood to be a fundamental truth. Peters' world might just about be OK if all the most important things in life were properly measured. But what happens if some or even all of the things that matter most can't be put into numbers? Then, if Peters is right, we are condemned to spend our time doing the wrong stuff. I write this as someone who spends a lot of time putting numbers to things. The important thing is to understand what numbers can *and can't* tell us. Some examples: it is essential to try to quantify carbon emissions, because the amount of carbon in the atmosphere determines the climate change we experience and that in turn will have a very great impact on human wellbeing. If we want to manage that carbon we need to understand the amount of it that is associated with everything we do. Otherwise our efforts can never be much better than random. On the other hand you can try quantifying how good your holiday was by counting how many sights you saw, hills you climbed or whatever, but it doesn't mean you will have appreciated any of it.

Metrics are ways of simplifying our world view, sometimes for better and sometimes for worse. All metrics are harmful if you give them too much power. GDP is no different. It is not a measure of human progress. Its simplicity makes it a tempting crutch for any politician who is feeling freaked out by the complexities of running a country full of real sentient people and is looking for a way to reduce their anxiety. There is information in GDP but no simple link to good and bad.

Learning from call centre metrics disasters

Twenty years ago, when call centres were a newish phenomenon, I spent some time training some of their managers in Sheffield. Their biggest problem was that staff kept leaving. After just a few years they were getting to the point where they had burned a hole through the population of the city; everyone who might possibly apply for a job with them had already worked for them, hated it and left. The mass of quantitative information that managers had at their fingertips was a big part of the problem. They knew how many calls each person made, how long they lasted and how many minutes it took them to get a sale. Most strikingly, they knew how long people were spending on their 'comfort breaks'. All this was driving the performance management. If I was challenged on how long I spent on the loo, I too would have wanted to leave. The message here is that by directing our attention in unhelpful ways, some metrics actually do us more harm than good.

What metrics do we need to take more note of?

Human and planetary health statistics, from carbon emissions to life expectancy and everything else along the way; biodiversity, pollution, human health, nutrition intakes and availabilities, and so on. These things directly matter and can be measured. Extraction of every material resource and all inputs into the environment from human activities.

We also need better wellbeing metrics. Sometimes it is better to do a bad job of measuring something important than a good job of measuring something irrelevant. Wellbeing can be sensed at the individual level, but all other attempts to measure it have to be treated with great caution. It is a doable exercise, provided the results are kept in context and taken with a pinch of salt. The Happy Planet Index, for example, combines societal measures and happiness with sustainability metrics.[5] Surrogate measures include surveys of reported happiness and levels of trust, measures of physical and mental health, suicide, crime, prison statistics (see page 175).

What metrics do we need to downgrade?

GDP and the length of comfort breaks for a start.

When a measure directs our attention one way, it has to be taking it away from something else. All quantitative metrics are simplifications of the world and therefore need keeping in context. They have an important place *alongside*, but not instead of, our rich qualitative understanding of the world and the human experience within it. All targets attached to metrics can be met in perverse ways. You could reduce prison populations by killing people or failing to notice crimes. You could increase GDP by creating a world in which no one does anything for free and out of kindness. National exam result statistics can be improved by pressurising kids into narrow-minded thought patterns which, as we will see in a few pages' time, are the opposite of the thinking skills we need to develop in the Anthropocene. We could increase life expectancy by keeping old people alive beyond their own wishes. And we can certainly make ourselves miserable by obsessing over money statistics.

It isn't that GDP can't serve a purpose. It can tell us something, but we need to learn to keep it in its place and remember that growth might not always be good. Comfort breaks, of course, represent a host of nerdy measures for which the data have suddenly become available and which, while theoretically useful, are more likely to warp our perspective on the world than do us any good.

In contrast, greenhouse gas emissions and other direct measures of the burden humans put on the planet are as close as we can get to directly measuring things that matter in their own right.

I try not to side with any political colour any more than is strictly guided by the evidence. The solutions we are looking for need to transcend all the existing political spectrums. However, the requirements for global governance of some challenges and for new approaches to growth do have implications that we can't ignore. We look now at what the free market can and can't do for us in the Anthropocene.

Can the free market deal with Anthropocene challenges?

Often not. It can't solve problems in which individual interests don't align with collective interests. So global challenges require some global governance.

Challenging the free market does not mean reverting to the fully planned economy. The answer is a clever bit of both. There is no escaping the need for sufficient global governance to deal with global problems. Free markets simply aren't capable of choosing to leave fuel in the ground, or for that matter, rejecting an efficiency improvement even if it were deemed detrimental to human wellbeing. Steps like this require market *interventions*, like it or not. Anyone whose ideology dictates otherwise is simply wrong. When it comes to climate change, for example, we have seen how rebound effects are undermining the efforts of the leading few because those who cut their impact create more opportunity for those who don't care to create havoc. So the change we actually see is far less than we would get if we could simply add up the savings that result from all the good practice that is going on in the world. This is why global emissions continue to rise despite the fact that we all know of people, companies and countries that are cutting their carbon. Most climate policy makers still fail to get their heads around the uncomfortable reality that, because of rebound effects, national climate change pledges can't simply be added together and subtracted from the 'business as usual' carbon curve.

With the exception of criminal gangs, all of us, even the staunchest neoliberals, believe in some rules. It is already not permissible, for example, to blow up a factory or murder a workforce in order to gain competitive advantage. All I'm saying is that the rules need extending enough to deal with the climate emergency and a few other critical issues that market forces fail to reach. As well as some hard rules, we can employ incentives and taxes that push businesses in a few key healthy directions.

Which is better, the market economy or the planned economy?

Both are hopeless on their own.

Do you prefer having arms or legs? Would you rather have a head or a heart? Do humans need to breathe air or drink water?

The twentieth century demonstrated the inadequacy of the centrally planned economy to most people's satisfaction.[6] The state, even if it is well intentioned (which is hard to ensure) is incapable of making all the decisions at ground level. A degree of market forces are in everyone's interests, *provided* that market works in a healthy way. What I mean by that is that lots of experiments go on at once to find the best ways of doing things. The best ones thrive and others learn from them so that they can do better too. So for example several aid agencies might simultaneously compete, learn from each other and support each other. If one was more successful than others, those at the bottom would feel a mix of pleasure that there were even more effective outfits than theirs doing the work that needed doing, along with a spur to do a better job themselves. And if someone going for a top job in one of those organisations were to lose out to a better candidate they would be happy that an even better person existed to do the work. That would be a healthy market. One in which the best can rise to the top, while everyone helps everyone else to improve. This community-minded kind of market is very different from the greedy neoliberal ethic. In 1776, when Adam Smith published *The Wealth of Nations* and supposedly founded modern economics, one of his stipulations for a healthy market, often overlooked today, was the requirement for all players to be moral actors.

The free market is provably incapable of dealing with global issues that require global governance. In his 'Common Cause' report, Tom Crompton called them 'bigger than self' problems, which can't be addressed by individuals pursuing self-interest.[7] Not everyone has yet had the imagination to grasp the catastrophe that lies at the end of the neoliberal experiment, because the results haven't *yet* been viscerally demonstrated to *everybody*. I hope we won't have to get to that point.

We need the self-adjusting nature of markets and we need the global overview that global governance can provide. And the hybrid between the two needs to be administered along with a set of values that enable it to work. We will come to those later (see page 191).

Our quick look at the roles of markets takes us into questions of wealth distribution. How much does this need to be engineered and how much is it OK to let things take their natural course?

What is trickledown and why is it dangerous?

Trickledown is the idea that when the rich get richer, some of this seeps down to the poor and everyone gains. But it is flawed because *relative* wealth determines what you can buy in a free market.

'Trickledownists' say that inequality doesn't matter as long as the poor get richer. Not only is there more and more evidence emerging that trickledown simply doesn't work,[8] but a quick look at food markets will show us how harmful it can be. In a global market economy, agricultural produce can, generally speaking, find its way to whoever is prepared to buy it for the most money. If the rich have a lot more money than the poor then they will be happy to pay more for it just to satisfy whims, niceties and extravagancies than the poor can afford to pay even if desperate. Cereals can be used to make biofuels for people on the other side of the world rather than feeding local people. Land can be used in nutritionally inefficient ways to provide what rich people fancy having rather than what poor people actually need. Conversely as the gap narrows, the poor become more able to buy what they need. Trickledown, to be blunt, is a neoliberal self-deception at best,

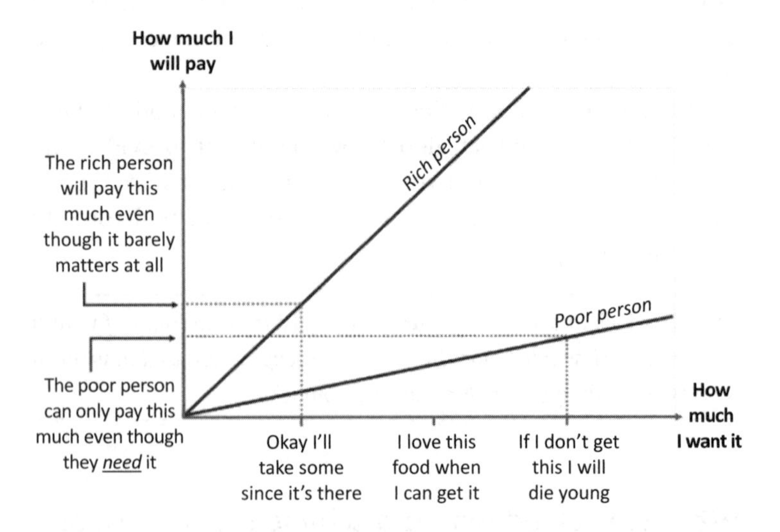

Figure 5.2 Disproving trickledown: a simple illustration demonstrating that if supply of a product is limited and inequality is high, the whims of the rich are met rather than the needs of the poor. This might apply, for example, to the output from a hectare of land, that could be used for cereals and pulses or for flowers and cattle feed.

and at worst a lie. The reason that some people go hungry is they don't have a high enough *proportion* of the world's wealth to buy what they need in a global market economy. Two solutions are to keep their food supply out of the global market (see the discussion about fish, page 36) or to narrow the wealth gap.

Why might wealth distribution matter more than ever?

Because people and countries that feel disgruntled won't want to participate in the global agreements that are required to solve the twenty-first century's biggest challenges.

Digging through the climate emergency agenda has taken us to the inescapable conclusion that the world needs to leave all but a small amount of the fuel in the ground. Once that has been grasped it doesn't take long to realise that this will mean the remaining fuel will need to be shared out somehow. And the same goes for all the renewable energy sources as well, even though, as we've seen, some countries have 200 times more sunlight per person than others.

Everyone is going to need to want our global arrangements to work because it will be easy for a disgruntled few to mess things up for everyone. This means we are going to need to be *fair*. This will require levels of trust and goodwill far above those we currently see within and between the nations of the world. There is simply no getting around it; we are all going to need to learn to treat everyone else with enough respect so that this kind of cooperation becomes possible. Too much inequality surely makes this impossible.

How is the world's wealth distributed?

America has 138 times more wealth per person than Africa. About half the world's wealth rests with 1% of the population, while the poorest 70% own just 2.7%.[9]

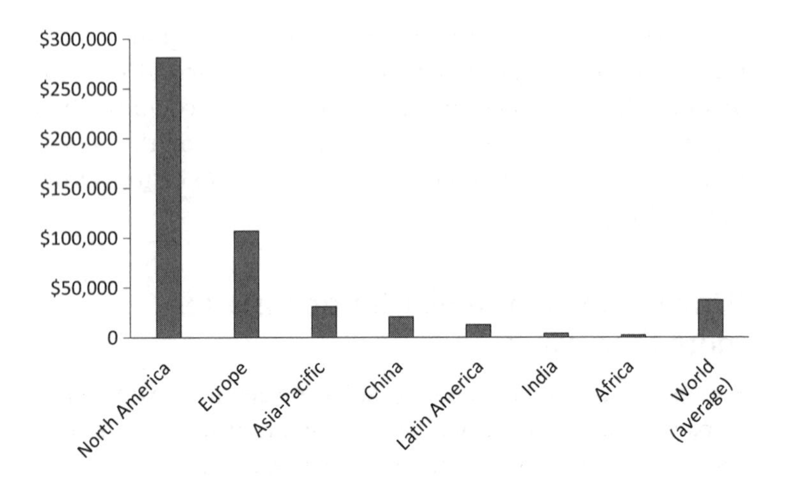

Figure 5.3 Average wealth per person in countries and regions.

In a sense, the wealth stats sum up humankind's track record on sharing. When we looked at what it would take for an effective global deal on the climate emergency, we came up against the question of sharing. Here we see how the world currently shares things out.

In terms of regions and countries, just over one third of the world's wealth is in North America, where 5% of the world population lives. Just under one third is in Europe, where about 10% of the population lives. About one tenth is in China which has almost a fifth (19%) of the global population. Less than 1% of the world's wealth is in Africa, where about one sixth (15%) of the world lives.

The richest 1% of adults own a million dollars or more while the poorest half own less than $10,000.[10]

Later on we will look at the values that can and can't enable us to deal with Anthropocene challenges. At the outset I laid down the simple principle that in my book everyone has inherent equal value in their humanity. In financial terms at least, these stats lay bare the gap between that value and the reality of global society.

Why are most Americans so much poorer than most Italians?

In the USA, so much of the wealth lies with so few that there is relatively little left over to go around the bulk of the population.

Unsurprisingly, as well as wealth gaps between countries and regions, there are also big differences in the way that money is shared out *within* countries. A simple gauge of this can be had by comparing average (mean) wealth with median wealth. The median wealth is the wealth of someone who is right in the middle of the population. In other words, if you stood everyone in a line from richest to poorest, they would be in the middle with as many people richer than them as poorer. The mean wealth is the amount of wealth everyone would have if things were shared out evenly. If this were the case (which by the way I am *not* actually advocating), then the median wealth would be exactly the same as the mean. The bigger the difference between them, the greater the inequality.

The USA is one of the richest countries in the world per capita, but most of its people are poorer than most people in many other countries that we think of as being poorer places to live. Examples include the UK, Norway, Denmark, Japan, Italy and even Spain. The median Italian is about twice as well off as the median American, despite Italy having only just over half of the wealth per person. And most people in Spain are wealthier than most people in America despite Spain having only about one third of the wealth per person as America.

The problem is that, in the USA, so much of the wealth is concentrated in the pockets of so few. If just the wealthiest 10 people in the USA shared out their riches among their fellow Americans it would be equal to a handout of about $2,000 per person (so $8,000 for a family of four). Alternatively, these 10 people alone could add almost a third to the wealth of every African citizen and more than quadruple the wealth of the poorest half of all Africans.[11]

When money piles up so high for so few, is it completely wasted? There is an abundance of evidence that making a very rich person richer does zero for their wellbeing. On the other hand, taking someone out of poverty is very likely to be extremely beneficial to their quality of life.[12] But despite this, my answer is that in theory at least, excess wealth is not *necessarily* always wasted. It all depends on whether the owners are good custodians. Does their wealth fund a planet-wrecking lifestyle of private jets, yachts and other crazy indulgencies, or is it all invested in community reforestation schemes? The decision to be wealthy can be seen as a decision to take responsibility for a greater part of the wellbeing of people and planet. I'm not talking here about a bit of high-profile philanthropy around the edges. I am talking about a committed and thoughtful attempt to direct *all* investment and spend in positive ways. We might ask how many of today's super-rich live up to this.

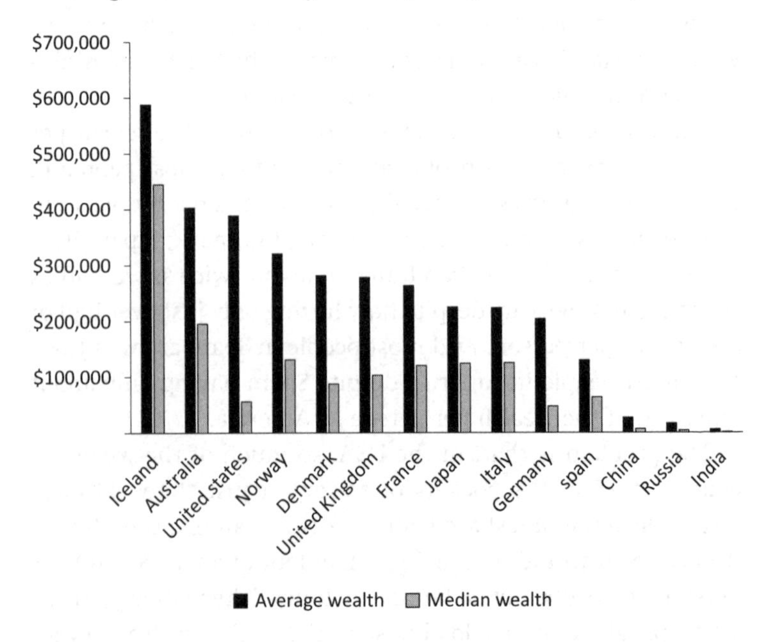

Figure 5.4 Average wealth and median wealth in selected countries. The closer they are, the more evenly wealth is distributed.

How has wealth distribution been changing?

Inequality has risen in most places in the twenty-first century. There are exceptions in Northern Europe. Equality has plummeted in China, the USA, UK and many European countries.

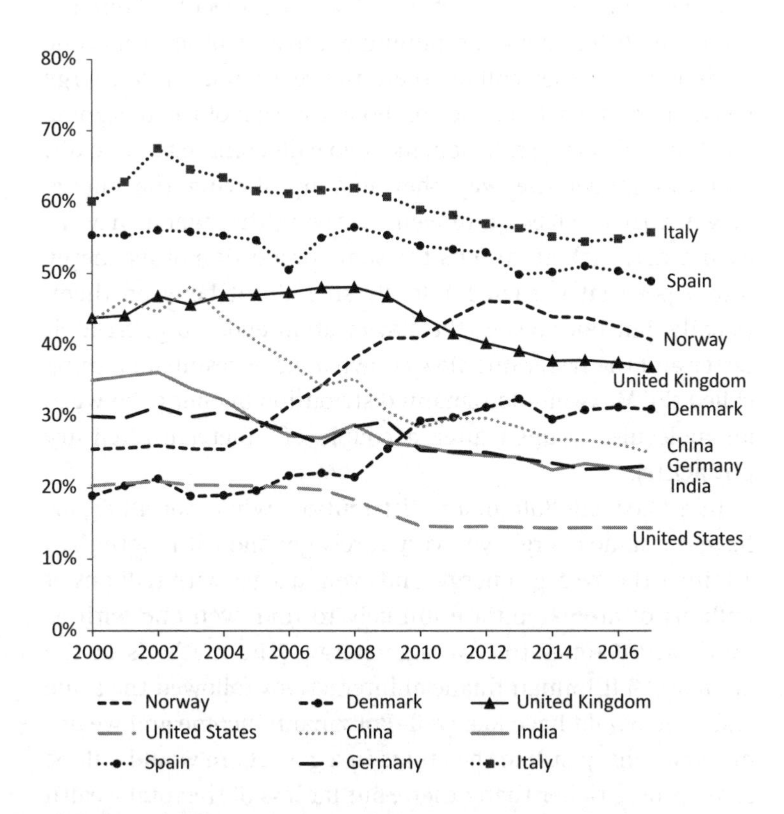

Figure 5.5 Median wealth as a proportion of mean wealth; a simple measure of how evenly wealth is shared out. The higher the percentage, the more equal: 100% represents total wealth equality. Most countries have got more unequal in the twenty-first century. Norway and Denmark are the exceptions shown. Italy and Spain are the most equal and the USA the least, having fallen to 15%.

When is wealth distributed like the energy in a gas? (And when is it not?)

Most of the time wealth distributes itself between people in a roughly similar way that the kinetic energy is distributed between atoms in a gas. But something different goes on for the very rich.

A gas can be thought of as a load of molecules bombing around at different speeds, bumping into each other from time to time. When they collide, there is a redistribution of energy between them that depends on how fast each of them is going and the angle of the collision, like two balls colliding in mid-air. You can model the way they end up sharing the energy between themselves quite well by saying that when they hit each other, each atom puts the same proportion of its energy into a pot that is then randomly shared out between them. Usually, but not always, the slower atom ends up going a bit faster and the faster one slows down a bit. It results in a thing called the Maxwell–Boltzmann distribution of energy between all molecules (named after a couple of nineteenth century physicists).

In a Maxwell–Boltzmann distribution, while not all atoms have the same energy, you very rarely get one with more than 10 times the average energy, and even in a gas with trillions of trillions of atoms you are unlikely to find even one with as much as 20 times the average energy. (The maths is in the endnote.[13]) If human financial interactions followed the same rules, we would have Maxwell–Boltzmann income and wealth distributions, just like the atoms in a gas. There would still be some people richer than others, but far less of the total wealth would be held by the richest few. The median wealth and median income would be a massive 79% of the mean. That is even higher than in Iceland, and quite a bit higher than Italy and Spain.

If the UK had a Maxwell–Boltzmann income distribution, the highest paid person would be on about £400,000 per year and there might be 100 or so people on more than £300,000. If the USA did the same, the highest paid would be on $900,000 and there would be about 100 people earning $650,000 or more. The mean wage in both countries would be unchanged but the median (the income of the person in the middle) would go up by about two and a half times. In other words, most people would be massively better off. There would still be inequality but, critically, just about everyone would be able to engage fully in society. The highest paid among us might be able to spend more time at the opera and top football matches, but nearly all of us would be able to go to them as a special treat. This level of income inequality feels about right to me. I don't have a problem with heads of states and the chief executives of the world's largest organisations earning 20 or 30 times more than a young adult in their first job, say. This feels entirely consistent with everyone being treated as if they have equal intrinsic worth as human beings. The Maxwell–Boltzmann distribution of income and wealth does have the problem that a very small number of people don't have much at all, but it would be a small and inexpensive tweak to sort this out.

Interestingly, if you look at human wealth distribution, it is roughly like Maxwell–Boltzmann most of the time. In other words, for most people, our financial encounters exchange wealth in a similar way to how atoms share energy when they collide in that usually, but not always, the rich spend money on the services of the less rich, resulting in the gap getting smaller.

BUT it looks as if the fundamental dynamics change at the top end of the wealth distribution. When it comes to the very rich, money starts to act like a magnet in a way that doesn't happen most of the time. Wealth starts to beget wealth. The rules of the game alter somehow and the very rich start to hoover up yet more wealth from the less rich. The central problem might be that the very rich are able to lend money at interest to those who don't have enough to get by. In this way they get to acquire money, not by earning but simply because they already have a lot of it. When this happens, the rich suddenly develop a tendency not to move back to the middle ground but to extend their lead, while the poor are pushed further down. The result is that a few people get to be many thousands of times richer than average, and most of the rest pay the price.

How can human wealth become more like the energy in a gas?

I'm not necessarily prescribing the Maxwell–Boltzmann as the ideal distribution for human wealth, but certainly the way we currently deviate from it at the top end looks unhealthy. So after some head-scratching and discussion, I have arrived at the following thoughts on how governments, businesses and individuals might help to take human income and wealth closer to a Maxwell–Boltzmann distribution. Some of what follows runs counter to normal advice and may feel challenging. They are emphatically not intended to be party political. They are just thoughts that I think flow from the logic of the argument. Don't shoot me if you think I'm wrong – just come up with some better ideas.

What can governments do?

Intervene to interrupt the many ways in which money begets money and poverty begets poverty; the mechanisms by which the rich hoover up more wealth simply by being already rich. This is the key unhealthy difference between the dynamics of human wealth and the energy of atoms in a gas.

- Enable universal access to high-quality education, health care and all the other basic necessities of a thriving life and career.
- Change the nature of inheritance so that it becomes a more evenly distributed passing on of wealth from one generation to the next.

- Find ways to discourage or prevent all forms of high-interest domestic debt and lending.
- A 'citizen's wage' so that everyone is guaranteed the means to live (see page 174).
- Transparency over pay, perhaps follow the Norwegian example in which everyone's pay can be looked up online.
- Finally, but as a last resort, adjust the income tax system until the distribution of take-home income becomes healthy.

What can businesses do?

- Make sure their salary distributions are roughly in line with the Maxwell–Boltzmann distribution.
- Make sure products and services are life-enhancing and are not exploitative of the less well off.

What can I do?

- Don't seek excessive profit from lending or investing. It might be reasonable to seek interest at the rate of inflation, and even to cover risk, and perhaps *even* to earn a sensible living from the service of fixing up the arrangement. But no more.
- Don't borrow at interest beyond inflation unless you have to. In other words, do not support money lending as an industry. One exception might be made for mortgages – since the rise in property price can be expected to roughly offset the interest paid and because the likely alternative, renting, is usually also a form of borrowing at interest.
- Don't gamble unless you want to become poorer. Not ever. Do not do so much as buy a lottery ticket. Don't be fooled by the 'good causes' con.[14] If you want to give to a good cause, it is much more efficient to do so directly rather than feeding the gambling industry a 5% cut and also allowing the government a particularly regressive 12% tax. The allowable exceptions to the no-gambling guideline are

schemes in which there is no third-party creaming off a cut other than a good cause that you want to support. In other words, the school raffle is OK!

- Do not seek wealth beyond your own lifetime needs and that required to raise your children and to look after the generations above you. The question of what level it would be right to set for those needs is, of course, a personal one for us all.

The UK gambling industry

On the 17th of May 2018 the UK government announced new regulations to reduce the maximum bet that can be placed on fixed-odds betting terminals (FOBTs – those are the modern equivalent of the one-armed bandit) from £100 every 20 seconds to just £2. The gambling industry is complaining that this will cut jobs and is telling the government they will lose £350 million of the £400 million per year that it gets from FOBTs in tax.[15] But in return gamblers themselves stand to save about £1.5 billion of the £1.8 billion they currently lose on these machines. Most of the revenue from gambling doesn't go to ordinary workers in the industry, but to the few very rich at the top.

In total, UK gamblers lose a massive £13.8 billion per year. The government takes £2.6 billion of this in tax, leaving £11 billion to pay for the jobs of 106,000 people who work in the industry, a few other running costs and the profits that are creamed off at the top. Jobs in the gambling industry fail my criteria badly and we are better off without them. In return for its tax revenue, the government acquires far more than £2.6 billion worth of social problems to deal with.[16]

STOP PRESS Very sadly, this legislation, which was originally intended to be rolled out within eight weeks was pushed back to 2020 following a deal between the Treasury and bookmakers.[17]

What should we invest in?

Every financial decision is an investment in one kind of future or another. This applies whether you are a government, a huge corporation, one of the super-rich, or just someone popping to the shops for some groceries. Every buying decision, on every scale, supports some supply chains and rejects others. Even bike maintenance is an investment in low carbon infrastructure. At the higher end of personal decisions, pensions and housing stand out. Pension portfolios now require scrutiny not just for the returns they offer but for the type of global future they support. Those signed up to employee schemes may feel powerless, but can still help by making their views known at work. Housing investments can support energy-efficient homes and sustainable urban design, rather than suburban sprawls of leaky homes that necessitate long commutes and high energy consumption.

For businesses and governments and other investors exactly the same principles apply. We desperately need investment in the following:

- renewable energy, especially solar, and the accompanying technologies to store and distribute electricity;
- transition to electrified transport;
- low-energy infrastructure from smart urban design to building efficiency improvements;
- population control through investment in the world's poor – specifically, education, contraception information and access, and ending hunger (see population questions coming up on pages 189–190);
- sustainable agricultural systems, including subsidising more people to work on the land, and more research into sustainable and truly biodiverse agricultural practices;
- carbon capture and storage; we need this despite the risks and despite not yet knowing how to do it properly;
- education to deliver twenty-first century thinking skills (see page 207);

- truthful media and democratic processes that can deliver intelligent decisions and global governance of global challenges;
- evolution in economics to meet the needs of both people and planet in the Anthropocene;
- support for people, countries and businesses that may otherwise be inescapably caught out by the speed of transition away from fossil fuels and yesterday's agriculture, infrastructure and culture.

In October 2018 the IPPC's report 'Global Warming of 1.5 °C' estimated the investment required to keep within that temperature limit to be $2.5 trillion per year – just 2.5% of world GDP.[18] What a bargain! Note however that this figure is more than 100 times greater than the hopelessly unrealistic average price of voluntary 'offsets' (see page 105).

How can these essential investments be funded?

Every divestment liberates an investment opportunity elsewhere. A carbon tax could also produce a massive fund.

In 2013, global investment in fossil fuels stood at over a trillion dollars while investment in renewables were just a couple of hundred billion. This is bonkers and none of us should be colluding in it. A trillion dollars for the investments I've listed would go a long way.

Global carbon dioxide emissions run at around 35 billion tonnes of carbon dioxide per year. Imagine, as a thought experiment, a carbon tax of $300 per tonne (equivalent to about a dollar per litre of car fuel) applied to all emissions. The funds raised would be over $10 trillion per year. We could use half to compensate those who unavoidably took a hit from the tax, perhaps redistributing wealth a bit while we are at it, and that still leaves plenty for the items in my list. Clearly quite a bit more detail needs working through, but the broad concept has passed the 'back of an envelope test' with flying colours. Jobs and pension funds would realign, but with no net loss to either.

What can fund managers do?

Ensure that value is understood as more than money. Define 'fiduciary duty' to include looking after people and planet. Treat carbon in assets as a liability and a risk. Divest and invest as outlined above.

In the two years since the launch of the first edition of this book, I have been increasingly involved with the asset management community. I've been encouraged by how many senior people in this industry are seriously asking what it will take to get the world's trillions pushing for the change that we need to see. Here are five things I'd like every asset manager to have their heads around.

(1) Asset managers have a legal 'fiduciary duty' to maximise value for asset owners (such as pension holders and other investors). The default interpretation of this has become the maximisation of *financial* value, but this does not have to be the case. All it might take is a bit of case law in which investors demonstrate that they have lost value, as they would define it, by not having environmental responsibilities looked after – this could change the default interpretation of 'fiduciary duty' to include both environmental and social values. We need to create an investment world in which it is normal to assume that asset owners care about their world, the people in it and their own grandchildren, and therefore that to not look after these interests properly constitutes suable negligence on the part of an asset manager. Such a shift will have enormously positive implications.

(2) Carbon in an asset portfolio represents both a responsibility and a risk. And when we talk about carbon it is essential to include all the carbon in the supply chains of an investment portfolio. I am encouraged to be able to write that some asset managers really are getting this now.

(3) Environmental and sustainability criteria need to have an important weighting in investment decision-making criteria

for all investment portfolios. And the quality with which these are scrutinised will be critical. (And we can expect that the public will ask increasingly searching questions.)

(4) Of course we need to divest from fossil fuel companies, and Bill Gates was wrong to say that this was not a vital action. There is currently still investment going to new fossil fuel exploration – of which the world needs absolutely zero – and to justify the bogus, disingenuous and frankly dishonest storylines that are still pouring out of the fossil fuel industry is just plain wrong. Just as the tobacco industry eventually moved on from denying that smoking caused cancer, so the fossil fuel industries have retreated to new lines of defence: extraction offset by carbon capture; investment in renewables alongside fossil fuel; claims that the renewables market is only ready for small-scale investments; and even – near where I live – a 'carbon neutral coal mine'! Yes, you read that correctly.[19] One argument for remaining invested in fossil fuel industries is that provided you have voting rights you can influence from within. The problem is that in order to apply any leverage you would need to be prepared to divest if the fossil fuel company was not doing the right thing. And none of them are.[20]

(5) Gaps need to be filled in existing investment so that the growing number of people and organisations that already want to know there is no fossil fuel in their pension or other investments can easily find alternative products to invest in. I know from experience that it is hard for a small business to find a pension scheme that is fossil fuel free and I know there are plenty of businesses looking for them (we found two in the end but it was a tough process[21]).

Why does the right tax make us better off?

Tax can be used to reduce antisocial behaviours, to fund things that make life better but which markets can't reach, and to control the wealth gradient.

Overall, tax doesn't make people poorer. That is just the direct impact that we can most clearly see. Tax does three key things. Firstly, it disincentivises some activities. Secondly, it raises money that can fund things that make life better. Thirdly, it can be used to change the way wealth is distributed. We have already seen that a totally free market can't deal with the challenges of today's world, and every other way of running the economy needs, and should welcome, the concept of tax. Tax is the reason we have roads, hospitals and governments. And it is a key mechanism to steer the low carbon world.

A carbon tax of hundreds of dollars per tonne will disincentivise the high carbon world and simultaneously fund the low carbon world. Overall it will make most people feel richer because it will all be paid out again and it will encourage us to make better use of our resources. Your gallon of car fuel will get more expensive but overall the electric car world will become cheaper. Under a carbon tax, those who manage their carbon carefully will be richer because they will pay less tax themselves while benefitting as much as everyone else from the things the carbon tax is spent on.

One argument against taxing the rich at a higher rate than the poor is that it disincentivises hard work. This is not true. It reduces the *financial* motivation for hard work. It reduces the motivation to do things that aren't fundamentally rewarding for other reasons – such as being useful, fun or otherwise meaningful.

> ***'Whenever you increase extrinsic motivation you always decrease intrinsic motivation.'***

In other words, the more you pay someone to do something, the less they do it for its own sake. What we need more of is people doing the jobs that need doing and are intrinsically valuable. It is a lie that the highest salaries are needed to recruit the smartest people. That strategy recruits smart, *money-driven* people, which is very different. Moderate money is better for recruiting people with more balanced motivations.

The quote is one I remember from Lew Hardy from over 10 years ago. Lew is a professor of sports psychology at Bangor University, a former chair of the UK's Olympic Psychology Committee, an international mountain guide and an all-round talented person. He used to do a bit of management training on the side and that is how we met. We had a gig together in which I wrote business games and he used them to help managers understand human motivation. My role was to set things up so that everyone had seemingly competing interests, hideous time pressure and a motivationally challenging stack of Excel spreadsheets to grapple with. His job was to help them understand how to manage their own and each other's motivations, backed up by some dazzlingly simple theory that I think has messages for us here too. (See this endnote.[22])

Tax is undoubtedly a massively important mechanism for reducing income inequality and countries use it to very different extents. The chart shows income inequality in selected countries before and after income tax. The inequality measure shown here is a well-established Gini coefficient, which ranges from 0% (everyone having the same income) to 100% (one person has all the country's income).

Ireland and Germany's income tax regimes take them from among the world's most income-unequal countries to among the more equal. The UK's tax system takes it out of the inequality doghouse to a somewhat more respectable position. Taiwan, interestingly, has similar income equality after tax to Ireland, but gets there without much recourse to income tax. Brazil and Peru, for example, have high income inequality before tax, don't use tax to sort the problem out and are left with hugely unequal societies. (For more stats and data on tax and income inequality follow this endnote.[23]) For all the moaning that goes on about income tax, most people are far better off for it, in relative terms at least, and in absolute terms too, provided the money is tolerably well spent on things we can all benefit from.

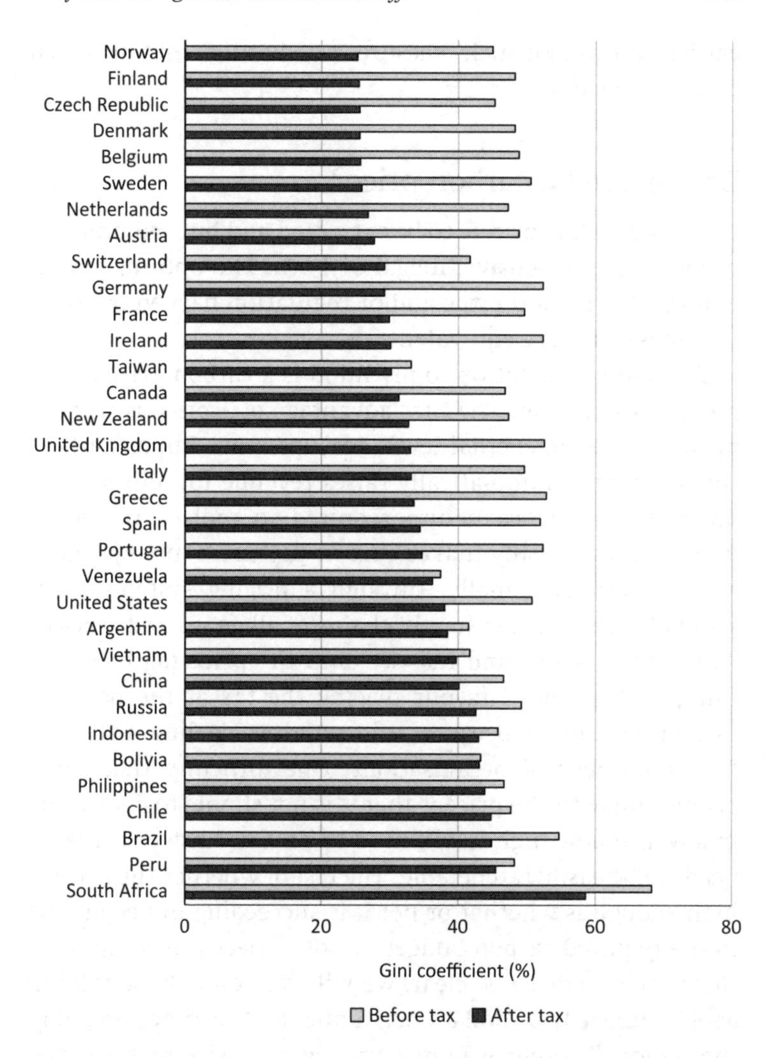

Figure 5.6 The Gini coefficient of income inequality before and after tax for selected countries in 2015.

Stewart Wallis, former head of the New Economics Foundation and now co-founder of WeAll,[24] makes the interesting point that taxing the rich might be thought of as a line of last resort and it is even better to start redressing the gap earlier on through such measures as education and universal child care facilities.

Do we need a carbon price?

Fossil fuel will continue to be extracted and burned unless it becomes too expensive, illegal or both. The options are an enforceable carbon price and/or regulation backed by fines. In a sense they are equivalent.

The simplest solution to my mind is a carbon price at the point of extraction. The first advantage of a price is its simplicity at the conceptual level. Perhaps more importantly is the fact that it automatically raises revenue for all the technologies and infrastructure required to replace the fossil fuel. It automatically makes carbon capture a money generating enterprise. Finally, through a pricing system, every scrap of carbon in the supply chains of all goods and services will automatically find its way onto the price tag of everything that is sold. I favour placing the tax at the point of extraction simply because this requires dealing with the fewest number of organisations. One difficulty that some people cite with the price is that it is not simple to determine in advance how high it will need to be. One answer is that it needs to be easily ratchetable. The test of whether the price is high enough is whether or not it is succeeding in keeping us to the required carbon budget. If not, it needs to go up until this is achieved. To be clear, we will very quickly be talking about hundreds of dollars per tonne, and in time, probably thousands. Another way of ensuring the price matches the budget is to 'cap and trade'; set the carbon, allocate rations, but allow these to be traded. The problem here is in defining the initial allocation.

A further note on a pricing structure is that it will also need to be smart enough to include other greenhouse gases, as well as deforestation. There will also need to be some important details in place so as to head off unintended consequences in advance. Perhaps the most obvious of these might be a dash for biofuel eating into the food supply.

Unless anyone has a better way of making it too expensive, illegal or otherwise not worthwhile to extract fossil fuel, we need a carbon price. So it doesn't matter how hard you think this is going to be to achieve, we may as well get focussed on the challenge. And somehow that price needs to be globally enforceable.

The difficulty in bringing it about is that it entails brokering the global deal of all deals. Quitting fossil fuel will have totally different implications for different countries and somehow that will need taking into account in such a way that everyone can sign up to it. In short, we will need to get our heads around the ***problem of sharing***. This takes us inescapably into the question of how humans treat each other on every scale. Gulp. We will look at this more later.

How expensive will carbon need to become?

The price needs to be enough that the fuel stops coming out of the ground in line with carbon targets. For perspective, a carbon price of $100 per tonne of carbon dioxide adds just three cents per mile to the cost of driving an oil-powered car and about eight cents per kilowatt hour onto the price of coal-powered electricity.

The only test of whether the price is high enough is whether or not the fuel is staying in the ground, in line with climate goals. It is as simple as that. My guess as to where it will end up, assuming we want to restrict warming to 2 °C or less? $1,000+ per tonne by 2050.

Microsoft has already adopted a carbon price that they use within the company to influence such things as company travel. I love the concept. The problem is that the last time I looked, it was set at a piffling $7 per tonne. I'm struggling to think of a Microsoft executive leaving her car at home for the sake of the 0.2 cents per mile that this translates to. However, the mechanism is in place and all that remains is to set a sensible price.

How should I spend my money?

Before you pay for anything at all, try to understand the supply chain.

Today I bought a copy of a newspaper which, in my view, supports truthful investigative journalism better than any other UK media outlet. The food section in this book (Chapter 1) contains enough guidance to inform most grocery decisions – perhaps not perfectly, but it's a lot better than nothing. (There is even more on food in my first book, *How Bad Are Bananas?*) The last computer I bought was from what I judge to be one of the more environmentally sound manufacturers, and when it broke I got it repaired, twice, by a local small company. I don't always get everything right, but the point I am trying to make is that there really is plenty to go at, just in everyday life, and all of us can fairly easily get to the point of being well enough informed that we are broadly pushing for a better world through our spending and non-spending practices.

As I write this, BT, for example, is starting to put carbon management into its procurement criteria and thereby improving the sustainability credentials of everything it sells.[25] The more discerning we consumers are about our product supply chains, the harder companies work to sort them out.

6 PEOPLE AND WORK

Does it all come down to population?

Human impact = Impact per person × Number of people

> **So the growing number of humans is clearly an important dimension to the Anthropocene challenge but not quite the single issue that some people frame it up to be.**

One billion reckless people would easily trash the place, while 15 billion careful people could snuggle up together and be fine. (Although, if everyone was careful there wouldn't be 15 billion in the first place.)

It is an encouraging fact that when countries get above a certain level of prosperity, their population growth almost invariably slows or stops.[1] Less encouraging is the fact that this is usually accompanied by a corresponding steepening in the impact that each person has. If a couple chooses not to have kids so that they can afford to fly off on more skiing holidays, the environment does not win.

In terms of the effect on the planet, it is not so much the number of people as their total combined impact. Several hundred Malawians have the same total carbon footprint of just one European or North American.

It is possible to draw many graphs showing rampant and accelerating growth, either properly exponential or at least roughly banana shaped, of different facets of human activities and impacts. Examples include energy use, GDP, extraction of just about every mineral, emissions and dumping of many pollutants. However, population growth turns out not to have this shape. There have been long periods when it has been very steady and periods of huge growth. Recently it has looked like a rising straight line and all the predictions are that it will top out at a good bit less than double today's 7.5 billion. If you live in the UK, Bangladesh, the Netherlands or Hong Kong, twice the number of people hanging around probably feels a bit oppressive, and while it isn't good at the global level, neither is it, in itself, a total disaster. Population pressure is not something that we can blame for all the environmental trouble that we are currently walking towards.

One of the challenges as we reduce global inequality is to help people come out of poverty without incurring the huge carbon footprints that are currently normal in richer parts of the world. They need to be able to leapfrog past the destructive lifestyles that are normal in richer parts of the world into the high-quality sustainable ways of living that, in my part of the world, more people are just beginning to aspire to. The better we role model the transition in the developed world, the cleaner the lives others will aspire to as they come out of poverty.

In rich countries, every additional baby tends to place more impact on the world than does a baby in a poor country. They usually use posher nappies, have more plastic toys, use more energy, eat more meat and so on. However, in poor countries population growth puts even more pressure on the local economy, very often adding to the stifling reliance on food imports. As things stand, North and South America between them export huge quantities of food. Europe more or less breaks even and most of the rest of the world relies on imports. In Africa, where population growth looks like being highest, this is set to be a huge problem. If there are 4 billion Africans in 2100, that continent will need a dramatically more productive agricultural system and/or enormous food imports.

What can I do to help with population?

What follows is not rocket science. I approach my top tips from a pragmatic point of view that I hope doesn't seem cold hearted.

They come with a respectful apology for any religious sensitivities I might be treading on. Finally, to help you put these top tips into context, I'd better also say that I write this as a middle-aged man who, having had two kids, deliberately made himself incapable of creating any more. These are my top tips.

(1) Please be careful not to cause a baby to be born unless you really want to look after it.
(2) Please don't encourage, pressurise or force anyone else to do so.
(3) Please try, wherever you have influence, to make it easy for others not to have babies unless they really want them.
(4) Push for greater investment in the poorest people. Specifically, push for
    (a) education, especially of women,[2]
    (b) information and access to contraception,
    (c) land reform so that more people have secure tenure of the land they work on,
    (d) end hunger – of course.[3]

And that's it. The more people who follow those tips, the better life will be for everyone who does get born.

One of the ways in which we distribute wealth is through employment and pay. But here in the Anthropocene, the whole concept of jobs could do with a return to first principles.

When is a 'job' a good thing?

When it is useful, fulfilling and appropriately paid.

Treating the total number of jobs as a measure of success is not useful because slaves or near-slaves would count while many people with purposeful, positive lives would not. And anyone who is paid to do things that pull society apart or trash the planet would also add to the job statistics, while others who work without pay to help hold people and planet together would not. Clearly the simple metrics of jobs and employment levels are unhelpful.

There are three reasons why a job can be a good thing.

(1) It does something useful in society. Examples include helping to make things that enable others to live well, jobs that enable society to function and thrive, or jobs that care for other people.

(2) It is fulfilling to do. Much of this might be because of reason (1), but it might also be rewarding for its own sake – fun, challenging and interesting perhaps.

(3) It provides a mechanism through which resources, and especially money, are appropriately allocated.

Sadly, many jobs don't fill all or any of these criteria. Nor are jobs the only ways of meeting the criteria. Useful things can be done by people without pay being necessary or appropriate. Fulfilling activities don't require payment either and money can be distributed through a myriad of payment and tax mechanisms. Here are some examples. If you work in a call centre, cold-calling people to try to persuade them to sue someone else for an accident they probably never even had, you probably hate your job, not least because you understand that its net impact on the world is negative. You and the rest of the world would be better off if we asked you not to go to work and paid you the same money anyway. Meanwhile neighbourly care and friendship isn't a job but meets the first two of my criteria.

In the UK I have recently heard both the gambling and arms industries defended on the grounds of the jobs that they provide. If more gambling or more weapons production can be justified as good things in their own right, then it is good to have people in jobs providing those goods and services. But if not, we need to find other things for those people to be doing, and in the interim it would be better to pay them to simply not go to work. Money spent on jobs for gambling or weapons can be spent instead on paying people to provide renewable energy, low carbon infrastructure, to grow food properly, enhance biodiversity, look after old people – or a million other things. The idea that we should have more weapons or more gambling simply or even partly in order to provide employment is fundamentally bogus.

How much of a person should come to work?

The whole lot.

I have spent a lot of time in businesses trying to have conversations about the things that matter most. I notice a big difference in the extent to which people feel able to express whatever they might find important, whether or not it relates to the current business targets, and whether or not it feels normal. You simply can't do the big-picture, multidimensional thinking that we need in the Anthropocene without using your whole brain. And that means taking it to work.

As I write this I am thinking back to a taxi ride after an unsatisfactory meeting with some of the top team of a company I'd been working with. I shared the ride with one of them. The day had ended up being all about money, and the word 'sustainability' had somehow been translated to mean 'bigger profits', and I'd felt caught out by world views that I hadn't seen coming.

As soon as we got into the cab, my travelling companion launched into an apologetic explanation. He was a full human being who cared about the environment, he told me, but at work he had to talk as if money was the only thing that mattered because that was how things were done. In other businesses, even some very large ones, I notice people having far more permission to express their whole selves.

What can I do?

Try to bring your whole self to work, including the bit of you that cares most for people and planet. Express yourself. Encourage others to do likewise. Anyone can open up the culture a crack but the more senior you are, the more important it is for you to do so, and the greater your responsibility.

Why would anyone work if they already had a citizen's wage?

Because people want to be useful, have purpose and make the most of their lives. Whether we think this will work probably depends on the confidence we have in each other.

By a citizen's wage, what I am referring to is an allocation of money that everyone gets, which allows them to live a healthy life and to participate in society. It might not buy tickets to the theatre or to football matches, but it would be enough to pay fuel bills, fund a heathy diet, a decent education and a carefully budgeted social life. It would take away the coercive pressure to work but not the positive innate human drive to do so. It would take away the fear of not working, but not the desire to be useful, to spend time in a meaningful way or even just to earn more money for life's treats. We would work because we wanted to, not because we had to. Employers would be obliged to make the working conditions acceptable. Boring jobs would have to go with higher wages. Voluntary work would be more possible for more people. Full-time hobbies likewise. You could jack in work to look after a loved one, to take up painting, farming, walking or anything you liked as long as it left you fulfilled. You could just chill out and live simply.

So why would anyone do anything? Because we understand that wellbeing entails being useful and purposeful. The effect would be to cut the *crap* out of the jobs market. With a citizen's wage, jobs that didn't score high enough on my three criteria simply wouldn't get done. In other words, we are talking about the true end of slavery.

Where would the money come from? It would come from tax. It is true that those with better paid jobs would fund those who earned no money. Access to a healthy diet and participation in society would become closer to universal rights. Those of us who are better off would benefit by living in a better society in which fewer people went to desperate measures to do useless or harmful activities just to get money. There would be fewer people doing jobs that don't need doing. The social gradient would become less steep.

Whether this model of the world would work or not depends on what view you take of human nature. Do we have an innate desire to be as useless as we can get away with, or to be useful, purposeful and creative? If you think it is the former then I hope you get a chance to enjoy the citizen's wage for as long as it takes for you to become restless and disprove your perspective through your own experience. If it is the latter, and there is plenty of psychological theory to back this up, then we should give it a try.[4]

In a way, the prison stats that follow are a bit of an aside from the main argument of the book, but I include them here for several reasons. Firstly, because they are both eye-opening and eye-watering; they give us one more reason to gasp and ask 'How can we humans be running the world so badly?' – and 'Surely we can do better!' Secondly, because they further illustrate both the inequality of the world and the mess we get into when we pursue the wrong metrics. And finally, because they further prepare the ground for the values discussion that is coming in a few pages' time.

What are my chances of being in prison?

It depends who you are. Pick a random US citizen and there is a 0.88% chance that they are in prison right now.[5] The average American will spend a staggering seven months of their life banged up. Things are a lot worse for men than for women and it is trickier living in some states than others. The average male in Louisiana can expect to spend one and a half years of their life behind bars. Prisons there are a profit-making venture. Sheriffs have financial incentives to keep the numbers up, and it all contributes to GDP. Things are a lot better for a girl born in Massachusetts who on average will spend only eight days behind bars. If you are black, the stats are vastly worse than for whites with American black men spending an average of about one day every three weeks in jail. Gulp! And a quick Google search of US prison images tells me that I would personally not like to spend a single night in one of them – ever.

Things are a good deal better in the UK, but not exactly great. About a quarter of a per cent of the male population is in prison but only about 0.01% of the female population. That is 10 weeks for the average UK male during their lifetime but only about three days for the average female.[6]

Norway is another story again. The average Norwegian can expect to spend 'just' three weeks of their life behind bars and, perhaps even more to the point, the experience of being there is vastly more humane. Why is the bang-up rate so low? For a start, compared to the USA and UK, people are far less likely to re-offend; just 20%[7] compared to 76% re-offending within five years in the USA[8] and a staggering 77% of UK prisoners re-offending within just the first year.[9]

Having a significant proportion of your population banged up in incredibly unpleasant environments has got to impact very badly on the average quality of life in a country. So what is Norway's secret? And what is the difference from the US approach? In the USA, prison is primarily a place for punishment, it seems. It is about society getting its own back. Both US and UK prisons appear to be, in large part, about revenge. It's an ugly word that few people like to openly use, but that's what is going on. After that prisons are supposedly deterrents and keeping people out of contact with the public. Rehabilitation is a pitiful afterthought.

In Norway, prisoners are human beings whose wellbeing matters just like everyone else's. Prisons are about repairing damage. Revenge, if we stop to think about it, is actually about multiplying the total damage. It's counter-intuitive for many that making prisons nicer can lead to less crime but anyone who struggles with this concept needs to spend some time thinking it through until they grasp the concept. The director of one of Norway's prisons, Are Hoidel, puts it like this: 'Every inmate in Norwegian prison is going back to the society. Do you want people who are angry – or people who are rehabilitated?'[10]

The mess of US and to a somewhat lesser extent UK prisons boils down to two issues. The first is the inequality that leads to some people being so much more at risk of going to prison than others. We have touched on nationality, gender and race, but education and wealth are also huge players. The second comes down to values and the question of whether the wellbeing of a prisoner is as inherently important for its own sake as anyone else's. We will turn to values in a few pages' time – and I hope that as you have been reading you have been feeling an ever more pressing need to explore this ground.

Although it is clear that we need to get better at evaluating things in non-financial terms, we have to notice also that Norway's radically more effective approach to prisons may in some way be related to its radically better and improving wealth distribution statistics compared to the USA (see the bar chart below).

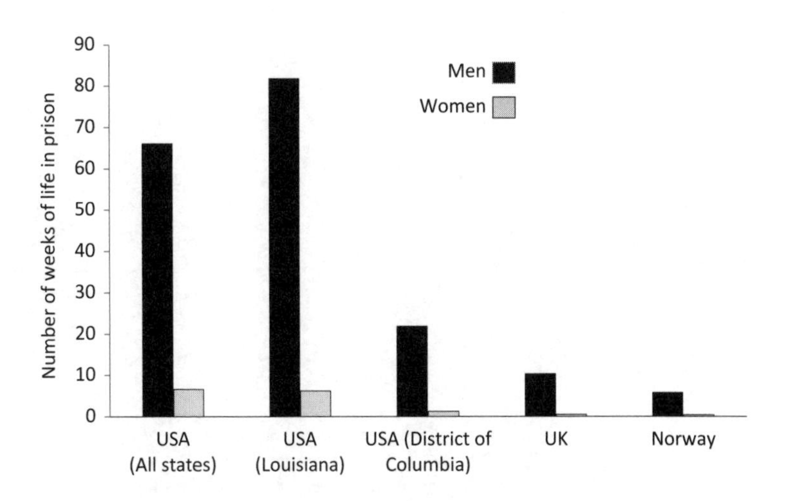

Figure 6.1 Number of weeks the average person can expect to spend in prison during their lifetime.

Prison statistics almost certainly give some indication of human misery and social problems. On the other hand, killing or releasing prisoners to improve the stats might not lead to an improvement in the wellbeing they are trying to track. I just write that to emphasise the need to keep all metrics in perspective.

Finally, getting back to money, for what it is worth, I'd better mention that prisons also cost a fortune. In the UK, the figure is something like £51,000 per inmate per year.[11] That excludes the costs of judicial and probation services not to mention the direct cost to society of increased crime through high recidivism rates or the social cost of the resulting fear and mistrust. In the USA things are a bit cheaper per head, averaging around $32,000 per prisoner per year in 2013, and taking the total annual spend to $74 billion.[12] In Norway, the cost per inmate is about three times higher than in the USA, but given the lower numbers, the burden on the Norwegian taxpayers is still several times lower.

7 BUSINESS AND TECHNOLOGY

Since so much of how humans do life is in need of rethinking, it is hardly surprising that as part of that we need to look again at the way we do business. This short chapter is not a comprehensive guide for business and technology in the Anthropocene but just has a few top-level thoughts, drawn partly from the evidence and logic of the book so far, and partly from 20 or more years consulting in organisations, including a few of the world's technology giants.

When is it good that an organisation exists?

When it does a useful job, is fulfilling to work for and enables appropriate distribution of wealth.

The criteria that I applied to jobs a few pages back also apply to every organisation. They should exist to meet three purposes. The first is the provision of useful and worthwhile goods and services – things that stand to enhance wellbeing of people and planet now and in the future. Secondly, they should provide meaningful and fulfilling ways for the workers to spend their days. Thirdly, they need to contribute to the appropriate distribution of wealth such that all people have the resources they need to have quality in their lives. Businesses that don't meet all these criteria should sort themselves out as a matter of priority and those that meet one or fewer should probably simply close down. If you work for an organisation that doesn't meet the criteria, and you don't want to be part of the problem, I think you need to leave or find an effective way of changing it from within.

Note that maximising shareholder revenue doesn't feature in the criteria at all. Maximising shareholder *value* potentially could, as long as the shareholders are operating from an Anthropocene-fit set of values that explicitly go beyond financial growth.

How can businesses think about the world?

Several of the businesses that I work with say that they want to 'leave the world better than we found it', using either those exact words or something similar. This sounds good, but here in the twenty-first century, it needs a bit of unpacking. What does this organisation think a better world looks like? What is the route by which it helps to bring that about? It isn't possible to answer these questions properly without taking time to stand right back from the problem. It requires a level of ***perspective*** that is rarely encouraged in the day-to-day commercial world. It can be scary to entertain any but the most banal answers. We've seen that much of the way we do things isn't fit for the Anthropocene, so if a business asks carefully and seriously what its role is in the world, the answers are very likely to shake its foundations at least a bit. Some courage is almost certainly going to be required.

Businesses need a vision for the world they are pushing for, and a coherent plan for how they help to bring that about. They need a ***systemic*** understanding of the full range of both their direct and their indirect impacts. By now, the next sentence may be too obvious to be worth writing, but just to be clear:

In the twenty-first century it is totally unhelpful to have organisations that exist primarily in order to make profit.

That is different from saying that they mustn't under any circumstance do so, but as a reason for existence, it is simply unfit for today's world. The profit motive has to be yesterday's thinking. If you work for an organisation like this, please challenge it and/or leave. If you feel you can't do either then you are a bonded labourer.

How can a business think systemically?

There are lots of ways of doing this. There are also many thousands of consultants eager to help if you don't want to do it all in-house. Some of them are very good, but I advise checking their world view carefully before engaging. Here is just one example of how things might be done in a simple way. It is based on how one of my clients started to think about climate change. (In a perfect world we would not have been thinking about the climate emergency in isolation from other global issues, but you have to start somewhere.)

We had got to the point of agreeing that in some form or other there needed to be an enforceable global deal to leave most of the world's fuel in the ground. Then they asked what it would take for that to be possible and concluded that political will was the most critical factor. So next they asked what it would take for the political will to be in place, and then what it would take for those things also to be in place. And so on. In this way they quickly built up a simple map of the critical factors for change. After a little bit of neatening up, this turned into the very simple diagram shown below. It wasn't intended to be definitive or perfect. They just needed something good enough that they could quickly relate to. Next, they asked where an organisation such as theirs could have influence – and it turned out to be just about everywhere since this was a big and powerful company. The final stage was to look at what they actually *wanted* to do given all those options.

There is no particular magic or rocket science in this process. Anything along similar lines will do. I've used a similar method with several other companies of very different sizes. The point I'm making is that systems thinking is not particularly hard; you just need to decide to do it.

A much more carefully designed and digital tool for helping organisations of every kind to think systemically has been designed by Bioregional and is framed around their 10 principles of One Planet Living. (See box.)

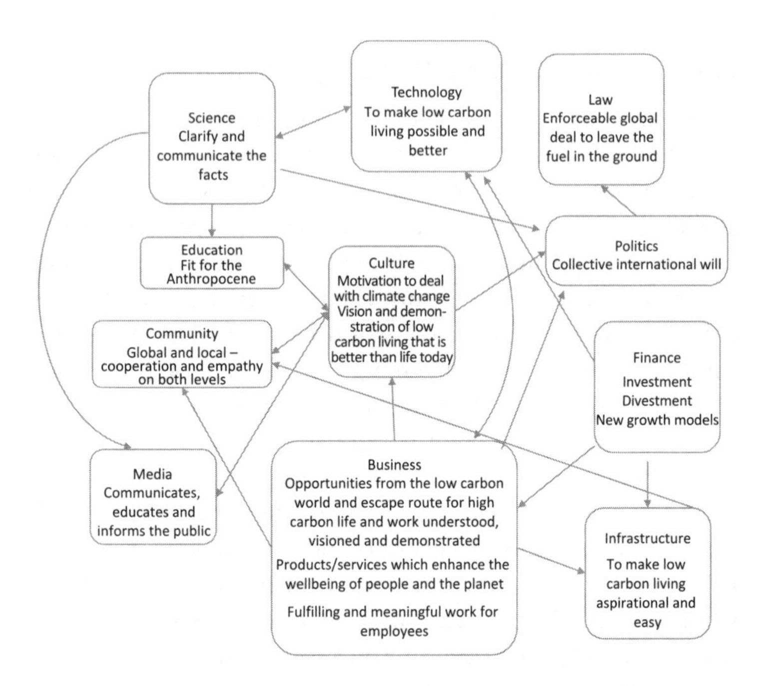

Figure 7.1 What will it take for the world to agree to leave the fuel in the ground? A simple example of systems thinking. This is one map, drawn up by a few people within one company in order to look at how they might be able to help.

Bioregional's systemic tools for One Planet Living®

At Bioregional they have been clear for over 20 years there is no planet B. They have been using their framework of 10 principles for One Planet Living with businesses and communities all over the world.[1] More recently, in recognition of the systemic nature of the Anthropocene challenge, they have developed a neat tool for mapping all outcomes, actions and indicators onto this framework in a non-linear and non-hierarchical way to produce visual maps of how things fit together, i.e. showing the interconnectedness of everything. Co-founder Pooran Desai explains that if you want people to start thinking in systemic ways, talking about it only gets you so far – you need to provide the tools for them to do so. Their oneplanet.com toolkit is free for companies, cities, other organisations and individuals to use.[2] If you like you can pay to add in other criteria on top of the One Planet principles, but when pushed, Pooran says that in his experience, company objectives usually either fit into one of the 10 principles, or conflict with them and therefore need to be challenged.

Bioregional's 10 principles for One Planet Living are:

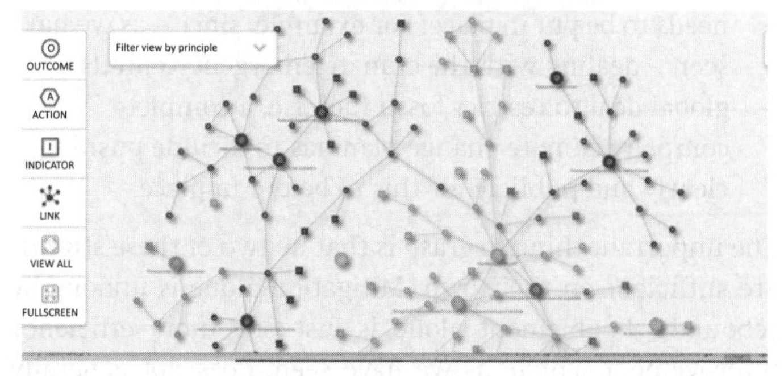

Screenshot from oneplanet.com's mapping of company actions and initiatives, which shows the interconnectedness of issues.

Three strands to an environmental strategy

The following model comes out of my work with three large tech companies, each of which would like to claim that they enable the low carbon world not only by cutting their own carbon, but by providing technologies that enable others to do likewise. Sadly, as we've seen, those efficiency-enabling technologies don't cut the world's carbon at all unless some global arrangements are in place. So the enablement story only works as part of a three-strand attack on carbon.

(1) **Improving own impact.** A company improves its own impacts on every environmental factor, in line with what the best science says needs to be happening. For example, with respect to the climate emergency, it cuts carbon emissions of goods and services from 'end to end' – manufacture, right through to use and end of life for the product.

(2) **Enabling others to improve their impact.** Its goods and services enable others to do life in ways that are better for the environment. As an example, an ICT company might enable energy savings for its customers or enable a reduction in flights.

(3) **Pushing for global arrangements where needed.** The company pushes for whatever global management needs to be put in place. For example, since – as we have seen – dealing with the climate emergency entails a global deal to restrict fossil fuel use, a complete company climate change plan has to include pushing clearly and publicly for this to be put in place.

The important thing to grasp is that no two of these strands are sufficient on their own. Mitigation alone is undone by rebounds. Enablement alone is just one more efficiency improvement which, as we have seen, does not generally lead to better global stewardship, unless there is a global constraint on the total environmental burden. Enablement

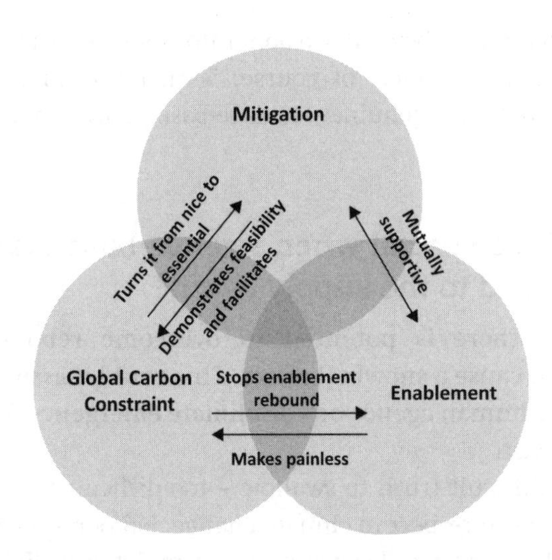

and mitigation are both methods by which **the global constraints that we need** are made possible.

What is a science-based target?

A target to do what the science says needs to happen in order to look after the world. Obviously, all environmental targets should be science-based.

We have seen that there are practical challenges that the world needs to address. A simple, essential principle is that the business response needs to be coherent with what the science tells us needs to be happening in the world to meet those challenges. With respect to the climate emergency, as I write this, there happens to be a very healthy and, if they can get the details right, potentially, quite important initiative that is gathering momentum and that has exactly that name; the Science Based Targets initiative (SBTi).[3] But the same principle applies to any environmental or health issue; biodiversity, antibiotic resistance, pandemic exposure, and pollutions of every kind including plastics, and indeed the SBTi is beginning to extend its range beyond carbon into some of these areas. Of critical importance, of course, is that the SBTi actually has up-to-date and genuinely science-based criteria for approving targets.

What is so special when science-based targets are applied to the supply chain?

Suddenly there is potential to overcome rebound and instead to cause a snowball effect. This could massively help to deliver human agency on the climate emergency. Worth a shot at least!

It is a difficult truth to swallow – too difficult for many so far – that with respect to climate change, balloon squeezing or 'rebound' effects have been almost or completely nullifying the best actions of people, companies and countries to date. (See Appendix.) Science-based targets (SBTs) are currently mainly adopted for scope 1 and 2 emissions only (i.e. covering direct emissions from a company plus those from the generation of the electricity that it uses). Sadly, these on their own are completely undone by rebounds. Unless every company adopts and meets scope 1 and 2 SBTs, those who don't play the game will take up the slack, generated by those who do. The emissions simply migrate to elsewhere.

However, if companies apply the science-based target approach not just to their energy use but to their entire supply chain, then there is potential for a totally new dynamic to kick in. I'm talking about companies signing up to ensuring that not only are their own emissions in check, but so is everything that goes on in their supply chain. The targets need to be met through robust incentives in the procurement contracts with suppliers. With these in place, the whole thing develops snowball characteristics because their suppliers now have to sort out *their* carbon as well, or face a competitive handicap. And so in turn do the suppliers' suppliers. I am not saying this is easy. But in principle at least, it has potential as a genuine mechanism of change.

There is another related strength to this approach. If a country were to decide not to honour the letter or spirit of the Paris climate agreement, then companies within that country might independently want to say 'we are still in'. But the question remains as to how they can make 'Paris compliance' real when they are part of a country whose emissions are going in the wrong direction, and which, for example, will probably be ending up with a high carbon electricity grid. With robust supply chain science-based targets in place, such a company can meaningfully be said to have detached itself from the rogue actions of its government and placed itself in a position in which other responsible businesses around the world can be relaxed about trading with it. Furthermore, individuals like you and I will be able to buy their products without breaching the Paris compliance of our own lifestyles. At the time of writing, I'm talking, of course, about the businesses which operate within the US economy.

UPDATE. I am so happy to be able to write that in the autumn of 2019 Microsoft announced supply chain targets compatible with a 1.5 °C world. Actually, BT had led the way in their sector with supply chain science-based targets a year or two earlier, although they did not at the time feel able go as far as a 1.5 °C trajectory in a world that was not yet universally calling for better than a 2 °C pathway. Now they have done so. And it looks like huge chunks of the information and communications technology (ICT) sector might be following BT and Microsoft's examples. The snowball effect might just be happening.[4]

Especially in the high-tech end of the corporate world, many people are tempted to think, or simply assume, that technology will be the primary 'force for good' that can guide us to the sustainable world. But since we would not have been able to get ourselves into the Anthropocene without it, this notion deserves much closer examination.

Do we drive technology growth, or does it drive us?

Right now we are slaves to a trajectory, but that doesn't prove we have to be from now on.

The late professor Stephen Hawking owed the last two thirds of his life to modern medicine. Without some of the advances of the past 100 years I too would be dead several times over, and several of my friends likewise. We are all grateful for the technology that keeps us going. But for all its advantages, Hawking also sensed danger ahead: '*Now, however, technology has advanced at such a pace that this aggression may destroy us all by nuclear or biological war. We need to control this inherited instinct by our logic and reason. We need to be quicker to identify such threats and act before they get out of control.*'[5]

Because we invent our technology, we talk about it as something we surely must be in control of, but is this really right?

Suppose a new technology brings an efficiency improvement. The first to adopt it gets a competitive advantage. Before long, to stay in the game everyone has to use it – for better and worse – and the way we live and work has evolved a fraction. Soon another new cutting edge arrives and the process repeats itself. We get to live longer, travel further, communicate more. We get to have and do more of just about anything we want. That is how the dynamics of economic evolution have worked for millennia, and in this book we have looked at some of the rising curves that have come about as a result. Could we stop them if we wanted to, or are we trapped?

What if a technology were to lead to an efficiency improvement but actually got in the way of quality of life? Would it be *possible* not to have it? If I'm a sales rep and all my competitors have access to smart phones with emails 24/7, then I'm disadvantaged if my customers can't contact me at the weekend like they can my competitors. We all have to adopt it. We may or may not love it but it wasn't a choice. If a supermarket prefers to have humans at the check-out because of the social benefits they provide for both staff and customers, how long can they hold out against competitors who save money by replacing people with machines? Stephen Hawking described artificial intelligence (AI) as 'either the best or worst thing to happen to humanity',[5] while Elon Musk, one of the great entrepreneurs of the automated electric car, also looks on it as an existential threat. Whether or not the specific concern over AI is well founded, it is clear that technology as a whole has brought many good things over the millennia, but has now taken humankind to a dangerous place and we need to handle it in a radically different way.

Is this possible? I have already said that I am not a determinist. I don't think any of us can prove the existence of free will or otherwise, but I write this book as if there is such a thing. Most claims that our future is pre-determined look to me like ways of dodging the challenge of trying to make the changes we need to see. Taking control of technology, across the board, in the way that we now need to do, won't be easy. We will need to raise and change our game. But that fact neither detracts from the necessity nor proves the impossibility.

How can we take control of technology?

This is one of the defining challenges of our age. To do so, humanity will have to raise its game.

While technology has a vital role in helping us live well and deal with the challenges we face we also need to be clear that it is not the unquestionable force for good that it is currently often assumed to be. Technologies need to be assessed in a discerning way against criteria that matter to us. Those might not be simple, might be variously understood and might become the subject of hot debate, but the debate needs to be had with thoughtful seriousness, and its outcomes need to determine what we do and don't take forward.

The simple case history of chemical and biological weapons is enough to prove a principle that it is just about possible (most of the time) to hold back on a development, while the well-trodden examples of radar, Spitfires and nuclear bombs demonstrate that it is also often possible to bring a needed technology along very fast. But the work on cancers shows that sometimes, even when we want to, we can't have the instant breakthrough that we'd like.

We create the future we want through the things we invest in and spend money on. New technologies require research, development and deployment (R&D&D). Once the R&D is done, it is a lot harder to hold back on the second D. Better still if we can hold off from even doing the R in the first place. Some technologies need to be starved out while others need urgent nourishment. Governments clearly have a huge role in directing research and industry support into the right areas, but all of us can help with everything we spend money on and invest in.

We simply can't get away from the question of values. Whether we like it or not, they have popped up as critical factors in dealing with the challenges of energy, environment and food, and in how we do our economics and run our businesses. So, we turn to them next.

---

8 VALUES, TRUTH AND TRUST

All the pathways of this book seem to be converging inescapably on the question of values. It turns out to be the crunch point. I haven't manipulated it that way. In fact, if I could have avoided a values discussion, the book would have been simpler to write. I am conscious of being no more of an ethics professor than I am an economist. But the evidence is pointing towards some values that will help us live well in the Anthropocene, and others that won't. So, I write this section from a pragmatic perspective. I'm simply asking which values will and which values won't allow people and planet to thrive in the twenty-first century, and how we can end up with the right ones. Luckily, it turns out that our values are something we can actually shape if we want to.[1]

At the outset I made clear the values I was going to write from, and here I'll explain part of the reason why.

What is the evidence base to choose some values over others?

Earlier in the book we saw that the question of whether anyone will have to be malnourished in the future eventually comes down not so much to a discussion of technological challenges, or a debate over population limits, but to the staggeringly simple question of whether those of us who have plenty care sufficiently about those who don't have enough.

The exploration of energy and the climate emergency led us to the conclusion that global policy agreements are needed. We then found ourselves asking what kind of conditions would make it possible for the world to be able to agree essential things, like how to leave nearly all the fuel in the ground, and who in the world would get to burn the remaining carbon budget. We found ourselves looking at a necessity for fairness and universal respect.

Meanwhile the rising complexity of the technology, the interconnectedness of the practical challenges we face, and the escalating flow of analysis, opinion and downright fake news, mean that it is ever harder to get a clear view of what is going on. We need values that will allow us to avoid drowning in a confused sea of information and misinformation.

Finally, as human capacity for impact on our planet goes up, so do the consequences of carelessness. We've seen that we can no longer expand in the way that has been possible over the millennia until now. It is becoming increasingly easy for even a small maverick faction of humankind to cause havoc. We have seen the potential for human conflict to wreck the whole world in a way that never used to be possible. So, we've seen that we need to be capable of care and, sometimes at least, restraint. We need to be able to tell when enough of something is enough, and when that happens, we need to find a way of not wanting to have more.

One useful way of thinking about values is to categorise them into *extrinsic* or *intrinsic* values. There is extensive research to show that this categorisation works across many different cultures, however huge their other differences. Extrinsic values include money, power, status, image and material possessions. Intrinsic values include self-acceptance, awareness, connection to others, appreciation of and care for the world and everything in it, and the enjoyment of activities for their own sake. It is our values that motivate us.

There is a stack of research to show that people and societies that are more governed by intrinsic values and motivations end up being happier and healthier and treating the planet kindlier than those that are more governed by extrinsic values and motivations. Intrinsic values are associated with higher levels of personal wellbeing, lower levels of distress, depression and anxiety, acting in pro-social ways, sharing more, higher empathy, being less manipulative, and more positive ecological attitudes and behaviours. As well as the opposite of these, extrinsic values are also associated with narcissism, substance abuse, behaviours that harm others, higher inequality and a general state of unhappiness and stress. This correlation has been consistently found across many cultures and all ages of people.

What values do we need to be the new global cultural norms?

Clearly, we need to focus more strongly on all intrinsic values. But I want to emphasise three in particular that seem more essential now than ever. These are values that need to prevail across all cultures of the world. Quite apart from whether they are inherently 'nicer', we need them for purely practical reasons in order to survive in the Anthropocene. Cultural diversity is a great thing, just as long as all cultures share these characteristics. Any culture that doesn't is unfit for the twenty-first century, and we should all have a problem with it.

Here are the three values:

(1) All people are inherently equal in their humanity. With this comes the principle that all should be allowed, encouraged and enabled to live their lives in whatever way they find meaningful, provided this is *negotiated alongside the equal rights of others to do likewise*. (The principle, of course, includes but is not restricted to all questions of equality of race, gender, class, nationality, religion, sexuality and anything else you care to mention.)

(2) Respect and care for the world; its beauty, life-supporting complexity and all its life forms.

(3) Respect for truth – for its own sake. The honouring of facts, as far as they can be discerned. Allowing others to have the

clearest view of whatever you or they may deem to be evidence. Transparency over reasons, methods and personal interests.

All the solutions outlined in this book are dependent on these values. Humans won't thrive without them.

All the evidence and analysis tells us that for humans to thrive over the next hundred years and beyond, we are going to need to learn to be as respectful, as truthful and as kind to each other as we can. I could have said that on page one. Any of us could have said that, of course. I'm sure millions of people have said words to this effect before me. In fact, as I write, I realise that these three values tie closely with what Carl Rogers, the father of person-centred therapy, termed the three core conditions for the therapeutic relationship: *empathy*, *genuineness* and *unconditional positive regard*.[2] They are, of course as relevant to everyday life as to the therapeutic setting. Perhaps all I am really adding is that, here in the Anthropocene, these values are no longer a choice.

I'm probably no better at practising them than you are. But we all need to try to be better at it. Simple but hard.

Can we deliberately change our values?

The clearest evidence that our values are variable probably comes from examples of them going in the wrong direction. Neoliberals and free marketeers have been capable of moving us towards more individualistic mindsets through a variety of techniques; focus on money, establishing status around material possessions, advertisements that link happiness to material things even when those things are clearly pointless. We should take heart from these because if movement in the wrong direction is possible, the other direction is surely possible too. And we can learn from these negative moves about what might push us towards the values we do need.

There are also many studies to show that different experiences can trigger different values.

What makes our values change?

Our values move with the **messages we receive** and the **things we think about**. In Rwanda in the 1990s, it was sufficient simply to have one radio station repeatedly sending out the messages that some people were cockroaches in order to swing the tide of opinion with devastating consequences.

According to the theory, and it rings true with me, increasing extrinsic motivations reduces intrinsic motivations and vice versa. So, for example, paying people to do things reduces the drive to do those things for their intrinsic value. And focusing on the intrinsic value of a job reduces the need for extrinsic motivators.

Two things in particular push us towards extrinsic values: insecurities and materialistic social messages. If we are worried about not having enough money to look after ourselves or not being socially accepted, we are more likely to want more money and status. And if we are constantly receiving messages that our own worth and happiness is tied to our wealth and our possessions, those are the things we are likely to pursue. For example, one prevalent but poisonous message is that in order to get the best person for a job you have to pay the highest salary. It creates the fear that anyone who doesn't earn excessive money won't be seen to be valuable enough. The more the focus of a job is on pay and bonuses, the less the employee will be thinking about intrinsic reasons for doing a good and useful job.

Conversely when we feel secure that, regardless of our wealth, we will always have our basic material needs met and will always be socially included, we relax about possessions and status symbols. This requires social infrastructure and community (but emphatically does not entail tolerance of harmful behaviour). And if we are continually reminded of our intrinsic values, their role in our lives grows.

How to cultivate the values that we need

- Create mechanisms that reinforce intrinsic values. This includes unconditional access to health care and education, and other social support such as entitlement to adequate paid vacation, paid maternity leave and, quite possibly, a citizen's wage (see page 174). These structures will open up the possibility, for any of us, of living a materially simple life without fearing for our health, our kids or our old age.
- Reform prisons into humane environments that focus on rehabilitation (see the discussion about prisons, page 175).
- Focus on metrics that emphasise intrinsic values, for example downgrading GDP and upgrading wellbeing measures (see Chapter 5 – Growth, Money and Metrics).
- Create a business environment in which shareholder profits do not need to dominate and business objectives beyond profit can more easily take centre stage.
- Emphasise the importance of public and community service, perhaps through youth schemes and the employees' right to take community service days in addition to normal holidays.
- Find ways to curtail materialistic advertising, for example through restrictions on adverts aimed at children and the withdrawal of any subsidies for commercial adverts.
- Do not try to win an argument by appealing to unhelpful values. This might get you a short-term win, but overall it is an own goal. Examples include trying to persuade people to reduce their energy consumption in order to save money. While you might get some immediate behaviour change, more importantly you have strengthened the idea that all actions revolve around money. So, if it doesn't feel financially worth it to take the next step, there is less overall reason for doing so. In the same way, it doesn't work to sell environmental strategies to businesses purely on the basis of increased

product sales. The reason has to be, without embarrassment, that it is the right thing to do.

AT THE PERSONAL LEVEL

- Spend time thinking about the values we need, talking about them and reading about them. Try to develop communities that do this. Identify what they might mean for you in practice.
- Consume critically and mindfully. Try to identify and challenge the explicit and implicit messages and motivations behind adverts, films, news items and political arguments. Show our children how to do the same by asking questions like 'What do you think this advert is trying to make you believe?' Try to view fewer adverts. As my Buddhist friends would say, mindful consumption is not just about what we eat, drink and inhale, it extends to all the information and experiences that we consume.[3]
- Have experiences that bring you into contact with a wide range of people. Make personal contact with the people you feel most detached from and feel least empathy for.

We have already explored the first value in my list to some extent through the discussions of wealth and income distributions and even prison statistics. The second value has, I hope, pervaded most of the book so far.

Now, following an eruption of post-truth and fake news all over the world over the past few years, we turn to the third value; the critical questions of truth and trust.

If we are going to live well, humans need to raise their game in this area. The rising complexity of the issues and the rising flood of both information and misinformation, sometimes accidental and sometimes intentional, means that we need to get much better at cutting out the lies, disinformation and 'fake news'.

Both sides of the Atlantic are seeing worrying lurches in the wrong direction. The consequences will surely hit us before long, and I hope will give rise to a longer-lasting counter-swing in the right direction.

Is there even such a thing as 'truth' or 'facts'?

Perhaps skip this question if my using the words 'truth' and 'fact' doesn't trouble you.

I include this question and put the words in quotation marks simply because I'm fearful of armies of social scientists, post-modernists and any others who may be waiting to take issue with the concepts. I'm afraid I'm a simple practical guy and only have limited interest in this particular philosophical debate, so I dearly hope that a few carefully chosen words here can put it to bed. Here goes.

I do understand that facts may not exist and indeed that we may all only be figments of our own imaginations. However, I choose to live as if there is a concrete reality, complete with facts and truths, even though they are sometimes hard to discern or even approximate. For the record, I consider that the world is incredibly complex compared to our limited capacity for understanding it. Thank goodness. How boring if this wasn't the case! So, while it is a worthwhile exercise to try to create models for understanding how the world works, we should all be careful before privileging one framework of understanding over another. As such I'm uncomfortable with all forms of fundamentalism, and not least scientific fundamentalism. If any sociologists out there want a label to sum up my position, I think 'critical realism' is as good a term as any, although I use it with fear of opening up a hopeless linguistic debate and I recently had a chat with a sociologist who connotes all kinds of things with this phrase that don't apply to me. If you crave labels, then OK, call me a critical realist, but only as defined here.

To summarise, it is useful to think in terms of facts, even though they are hard to discern and all we ever get is a partial view. Anyone who wants to sink into a nihilistic vortex of relativism can do so without me joining them and it won't matter to them whether or not they read on.

Is 'truth' personal?

I want to be clear how I'm using this word because I have found that some people, especially in sociology departments, sometimes use it a bit differently. In my meaning of it, there is no such thing as one person's truth as distinct from another person's truth. If something is true, then it is a fact. Period. There is nothing subjective or personal about it. A person's *view* of the truth is a different thing altogether and always *is* personal. One person's view might be closer to the truth than another person's or they might simply be seeing it from a different angle. Or one of them might just be plain wrong.

So, if someone says, for example, that anthropogenic climate change isn't happening, it isn't helpful to say that this is their truth because it isn't any kind of truth at all. The fact is that they are wrong. They have a view of the truth that doesn't square with the evidence and is therefore false.

Often two people might be able to teach each other how to see more of the truth than either have managed on their own, precisely because they come at it from different angles. This can be hard for both parties, but given the need for richer understanding, the ability to do this is another skill we could all do with getting better at. If one party is wrong, there may be very interesting reasons why they hold their view and understanding that might be incredibly important.

Why is dedication to truth more important than ever?

Because the situation is more complex than ever.

There has always been plenty of misinformation in the world, by both accident and design. But now the complexity of the issues requires us to raise our standards and increase our intolerance of nonsense to a standard we have never seen before. Here in the Anthropocene, it is hard enough to work out what to do without journalists, politicians, activists or even friends disregarding facts. More than ever, we need the clearest view of the situation that we can possibly get and should not tolerate people or institutions that deliberately or carelessly give us otherwise.

If I could pick just a few things to take humanity beyond its inadequate response to the Anthropocene, one of them would be to create, better than ever before, a pervasive, everyday insistence on truth.

What is a culture of truth?

Quite simply, a world in which everyone knows that they are in career-stopping trouble if, within their professional life, a lie or deliberate attempt to create misunderstanding is exposed. Beyond this it is a culture in which credibility depends on the ability to acknowledge and be candid about all valid sides of an argument, neither being blind to one side, nor creating spurious controversy. Such a culture will enable the clarity of analysis and debate that we so desperately need and aren't yet getting. It will enable a properly intelligent approach to the Anthropocene to emerge. It is a culture in which the public can work out who to trust on, for example, analysis of the pros and cons of fracking or nuclear power or reducing greenhouse gas emissions.

Is it possible to have a more truthful culture?

If it has been possible to go backwards, it should also be possible to go forwards.

I think it is doable; not perfectly, but I think we can move in the right direction. Recent lurches the wrong way on both sides of the Atlantic prove that cultural insistence on truth is in fact a variable. I think everyone can help. I think it will take effort but what is required is fundamentally simple.

If you are a politician or a journalist, your role in helping to bring this about is pretty easy to work out. It is all very well for the rest of us to say that our ministers and senators, mainstream news channels, newspapers, bloggers, etc. should be doing a better job of defending the truth, but what can any one of us do to improve the situation?

What can I do to promote a culture of truth?

Actually, there is only one thing: *Insist on truth everywhere.*

But more specifically, here are four simple suggestions.

(1) Ask very carefully whether your sources of news are giving the best view of the truth that they are able to. If not, switch source, let them know why and tell all your friends about it.

(2) Consult several sources, not just one.

(3) Make sure all electoral candidates know how important truthfulness is to you. Let them know that you won't be voting for anyone who fails your standards, or if they all do, you will be voting for the most trustworthy.

(4) Raise the standards of honesty everywhere you go. Confront bullshit, rather than letting it go. Make it risky to propagate disingenuous nonsense. Perhaps try to do this with such a friendly smile on your face that you leave the other party room to change tack without excessive humiliation. (In some situations, this might also help you not to get beaten up.)

(5) Hardest of all, tell the truth yourself. Every one of us, if we are honest, finds this hard from time to time but it's got to be done, and sadly, there is simply no such thing as a white lie.

What can journalists do to promote truth?

Insist on truth. Here is what to do if there is strong evidence that someone you are interviewing has misled the public, either by lying or by seeking to create a false impression, or by being mistaken: pursue it hard until you either have an apology and a climb-down or a proper explanation or the person has left political office. Do not be distracted by other things on the agenda. If you know they have lied, misled or are clear-cut wrong, make it the only thing that matters until it is resolved.

Expose incongruence. If two strands of policy are incongruent with each other, expose the incompatibility and don't let go. Ask something like 'Is it that you hadn't understood the incongruence – in which case I expect you to go away and think about it and come back with new thinking and a clear change of position – or that you knew – in which case you are unfit for public office?'

Respect privacy. Of course, alongside this comes an obligation to keep our noses out of business that doesn't concern us. Not many of us would like to go on the radio with the expectation of having to be 100% candid about every aspect of our lives. With the expectation of truthfulness comes the right to privacy.

Respect those who change their mind. We actually need more of this. Open-mindedness is to be encouraged given the context in which no one has all the answers.

Don't present two sides of an argument equally if they are not equal. For example, some news channels did us a big disservice over the climate emergency for many years, and damaged their own reputations, by presenting fringe views that were mainly funded by vested interests as if they were on equal footing with almost all of the world's most respected scientists. Minority views propagated mainly by those with a vested interest need to be put into that context.

What can politicians do?

It's obvious but I'll spell it out anyway.

Be honest in both the letter and spirit of what you say. Do not stay silent if a colleague misleads – even if it supports your own position. It is depressing seeing politicians tolerating dishonesty in their colleagues just because they share the same political agenda. If you do that, you are part of it. If you take issue, you'll be respected by those who value truth.

How can I work out who and what to trust?

A culture of truthfulness deals with one side of the equation. The other side is for all of us to get smarter at working out who we can trust about what. Honesty is a big part of this but not the whole story. Too many of us too much of the time are placing too much trust in the wrong people, and the wrong media.

Here are six tests, the first four of which are adapted from management trainer Tim O'Connor's 'Keys to Performance' – a model he developed for assessing an individual's ability to get something done.

(1) **Are they competent?** Do you think they are smart enough to do the analysis? Do they have expertise in the areas you need them to have? For example, alarm bells might start to ring if anyone without scientific training starts casting doubt on a consensus held by the scientific community.

(2) **Have they had time, resources and access to information** to be able to understand the issues?

(3) **What are their motivations?** Does this person or organisation have a strategic or financial interest that could push them towards a particular position? Are they funded, directly or indirectly, by people, companies or governments that have a particular agenda? If so, ask very carefully indeed how much you trust the independence of the work. While I am talking about motivations, also ask about psychological drivers. Have they wedded their identity to a particular position that makes it hard to change their mind?

Clear examples of alarm-bell raisers include a media chain owned by a family with business interests and political alignments, a research institute that is funded by the fossil fuel industry or a journalist or academic who has used a controversial stance to gain a profile that they would otherwise have been unable to gain.

(4) **Self-awareness.** Do they take time to reflect on and understand their own emotional reactions, leanings and influences? This is important because without awareness,

these things drive us, like it or not. With awareness, they might still drive us, but at least we have more choice. Are they transparent about their own reactions to evidence and events? If a newspaper is left wing or right wing or any other colour, does it have a deep understanding of how that came about? If so, do you find their experience is persuasive to you as well?

(5) **Do they have a track record of being able to change their mind when the evidence has changed?** How readily have they owned up to their past mistakes? Is there evidence of them being able to evolve their position over time as their thinking matures. Here in the Anthropocene there is no one who knows everything about what to do. Anyone who isn't changing their mind from time to time isn't thinking adequately.

(6) **Have they ever sought to mislead the public?** I don't just mean have they ever lied. I mean have they ever sought to create a public view that is at odds with all the evidence that they know of. Examples include trying to make the public think that the NHS would have £350 million per week more if the UK left the EU. Or discouraging the electorate from listening to experts. Or sexing up a dossier to strengthen the case for war *even if* that person honestly believed it was the right course of action. Or trying to make the public believe, against clear evidence, that one crowd was larger than it actually was. If your information source has tried to do this, you can't trust them *unless* you have convincing evidence of rehabilitation since it happened.

What are some bad reasons for placing trust?

- **This person has a confident tone of voice.** People who practise twenty-first century thinking skills understand that the issues are complex and that they still have plenty to learn. They are feeling their way through the nuances in a way that might require hesitation.

- **This person never changes their mind.** Smart people are able to do so.
- **This person makes it sound simple.** While there is skill in pulling out simple important nuggets, distilling things down and creating models of understanding, overall, the issues we face aren't simple. For all of us there is a danger that we lunge for the comfort of simplistic answers.
- **This person *says* they care.** Look for evidence.
- **You have always read this media source and so have your parents.** The most popular papers in most countries have a poor record with the truth.
- **They dress smartly.** Some but not all of the best thinkers care about these things. And *many of the worst thinkers care about them far too much.*
- **Celebrity status.** This last point is surely too obvious to be worth writing down, you'd hope.

How can I tell whether to trust anything in this book?

This is exactly the question you should be asking . . . Tricky for me to answer so I'll let you run your own test, but here briefly are some things you might consider.

(1) Am I competent to write this book? I admit to being a generalist, but I would hope to have enough of a science background to be able to get the odd paper published in decent journals and to run a business that does leading work on corporate engagement with the climate emergency and to have two books behind me that have been endorsed by people, some of whom you might also trust. Large chunks of this book are my own research, much of it peer reviewed.

My career has wandered all over the place, and I hope that helps when it comes to writing about seeing things from a lot of different angles. Just about everything I've done feeds in somewhere. Here's a quick tour: Physics degree, Outward Bound tutor (to every kind of student you could think of), teacher, professional development trainer (from scaffolding apprentices to top managers), organisational development consultant (from Oxfam to Coca Cola and everything in between), one-time resident of a Zen Buddhist community, sustainability consultant (to tech giants, local community groups and most things in between), apple picker, call-centre worker (very briefly), product manager (for outdoor kit made in China, Wales and a few places in between), importer of fairly traded clothing, low-budget traveller, dad, cross-disciplinary academic, author. I'll stop there. I've done quite a lot of bits and bobs to feed into the mix.

(2) Time, resources and access to information. Five years on and off, but before that a great deal of mulling over, especially in the course of writing two previous books. It is true that I have often felt too thinly spread and weighed down with distractions. Plenty of others have helped, including a small army of interns and paid staff and many others who have offered their views – see Acknowledgements.

(3) Motivations. Yes, this book might help me make a living, but since book writing is usually a terrible way to earn money, I'm relatively free from financial vested interest. Egotistical motivations, perhaps. I'm no more immune to that than anyone else.

(4) Self-awareness. At least I'm going through this exercise.

(5) Track record of my ability to change my mind. Yes, I think I have a provable track record for this, for example having changed my tune over a few passages from my previous books.

(6) Have I ever sought to mislead the public? Not that I can think of and I don't think anyone has ever made this claim against me either.

9 THINKING SKILLS FOR TODAY'S WORLD

We have seen throughout the book so far how urgently we need to learn how to think in ways that let us deal more effectively with the situation we have created for ourselves. We need thinking skills and habits that fit the twenty-first century context of enormous human power and technology on a now-fragile planet. We've also seen the global interconnectedness of just about everything we do. It is not at all surprising that ways of thinking that have taken us to this place might not be the same ones that will help us to live well now that we have got here. The brain skills that we developed over the millennia as we expanded on a big, robust world are not the same as the ones that will let us do well on the small delicate spaceship on which we now find ourselves.

What new ways of thinking do we need in the twenty-first century?

Here are eight dimensions of thinking that we rapidly need to get better at. This isn't the full list of all types of brain activity that we need, but just the ones we urgently need to improve upon. Mine aren't the first and won't be the last words on the subject. This is just a list that I think comes from the evidence about the nature of the challenge as we've seen it so far, and what we have seen about how we need to respond. I'm not claiming any magic bullets, and I am just offering these eight areas up for consideration. I hope they give a good enough start point, rather than just an excuse to debate their shortcomings. However, I know there is room for improvement and if you have suggestions or a better list, please send them to me at Mike@TheresNoPlanetB.net.

By the way, in no way am I claiming to be a master of this skill set. But it is a wish list of the stuff that I think every one of us needs to cultivate as best we can.[1]

(1) **Big-picture perspective.** Since the problems we face are now global, our thinking needs to be global. In this book we have seen time and again how almost everything we buy and do requires a global supply chain and sends ripples around the world. We need to tune into this even though we rarely see any of it with our own eyes. We have also seen how the global system dynamics often ensure that small-scale positive actions in one place are undone elsewhere unless the global perspective is properly understood.

(2) **Global empathy.** A thousand years ago this skill might not have been needed at all. But now, our daily lives affect those of people on the other side of the world who we will never meet. This applies to all of us whether we like it or not. Since our circle of influence is global, our circle of concern needs to become likewise. This applies to all of us. Not many of us are used to this type of thinking. And, like it or not, they have influence over us too. If our sense of 'tribe' doesn't embrace the whole world, we are going to be in for a very nasty time. That doesn't mean there isn't still scope for feeling part of smaller tribes at the same time. There is still just as much place for family, community, place of work, country and even sports team. All these can still be part of our sense of who we are and what we belong to, but at the end of the day, all of us need to be able to keep in mind our shared and overarching global tribe. We have to get our heads and hearts around the idea that we are in this together because that is the only way *any of us* can live well. One possibly tempting but out-of-date idea is that we might get away, instead, with making sure that our own smaller tribe is the one that wins. The reason this traditional strategy is no longer viable is that as the world gets more fragile compared to our power, so the disharmony between people stands to have ever greater consequence.

Increasingly small numbers of people will be able to spoil everyone's party if they want to. That is the world we have created and we need to learn how to live in it.

What this means in practice is that a person being blown up in another part of the world needs to matter as much to us as it would if it happened in our street. A sweat shop 5,000 miles away needs to feel like having one in our own town. The repeated murder of children in schools in our own culture needs to hurt no more or less than the same type of event in a very different part of the world.

(3) **Future thinking.** The argument for tuning into a more distant future is the same as the one that tells us we need better global empathy. Our circle of concern needs to be the same as our circle of influence because otherwise that influence will be irresponsible and we will trash the place. To date, most policy making finds it hard to think far beyond the electoral cycle of four or five years and most people find it hard to get excited about events more than 40 years away. One possibility is that this might be because that is the length of a career. Another is that 40 years is about the average number of years that readers of this book think they have left to live. But our kids will need us to have cared further ahead than that. In 80 years' time climate change impacts will be many times stronger than they will be after just 40 years. I will be long dead but I hope my kids won't be. When I'm on my last legs, my kids will care about what my generation did. None of us want to be looking at a world that is falling apart, knowing that our kids know that we let it happen. Surely that thought alone, if we can contemplate it deeply enough, ought to be enough to tune us into the 80-year perspective. It's worth a try, surely.

(4) **Appreciation of the simple, small and local**. This skill is the art of slowing down, and savouring the people and things around us. This is learning a different strategy for achieving a sense of wonder about the world – to notice what we have already, rather than numbing ourselves with

ever-increasing overload of bigger, faster, newer and wilder stuff. This is the skill of simple gratitude. It is also an essential antidote to the great acceleration of which we are all currently part. This is the thinking skill that will allow us to be truly content not to grow the things that we know will harm us and every other sentient being. Since we can no longer, on the whole, expand our activities, we need to get better at appreciating what we already have in front of our noses. There is no point having more, buying more, doing more and flying further if we don't even notice any of it properly. This is the skill that will allow us to become satiated. This is the skill that will allow us to truly feel that 'enough can be enough'.

(5) **Self-reflection.** Closely linked to *appreciation*, this is the capacity to notice our own experience. It is about spotting how we are reacting, what emotions we are feeling and better understanding our own motivations so that we are not driven by them quite so blindly. Self-reflection lets us put ourselves in perspective. It is about understanding the world through our understanding of ourselves. It involves forgiving ourselves for our shortcomings so that we are better able to address them. Just like appreciation, this involves a bit of slowing down.

Self-reflection is also the first of two thinking skills relating to truth. It enables the humility and open-mindedness to notice when it is time to change our minds; when our views conflict with the evidence or when we are clinging, for whatever reason, to something that is no longer right. In a world where no one knows the answers and we are all feeling our way, the capacity to change our mind for good reason is a strength not a weakness, and we should applaud it, not least in our politicians.

(6) **Critical thinking.** The second of the truth skills, this is the capacity to make well-founded decisions about who and what to trust. It involves asking careful questions about what we are being told, and the motivations, values and

competencies that lie behind it. Coupled with self-reflection, which enables us to assess our own vulnerabilities and susceptibilities, critical thinking gives us the ability to discern fact from fiction in an increasingly complex media and political sea of claim and counter-claim. This is the skill of standing back to put everything we see and hear into its *context* and into *perspective*. The world's information flows cease to be divided so clearly into good and bad, but become nuanced. Coupled with the context of its motivations, this is the skill, more than any other, that can stem the tide of fake news. This skill compels our media and our politicians to raise their game.

(7) **Complex and complicated thinking.** Because we have created an ever more complicated AND complex world, our capacity for this kind of thinking simply has to rise in step with it. We've touched on the spiralling complexities of sorting out the energy mix in even one country. We have seen that neither climate change, nor energy security, nor feeding the world, nor any of the other presenting physical challenges can be tackled in isolation. We need to get our heads around the interdependencies at the same time as dealing with the growing technical challenges that lie within each small part of the puzzle.

(8) **Joined-up perspective.** If only technology, or natural science, or sociology, or philosophy, or theology, or politics, or art or literature alone could deal with the Anthropocene. They can't. Even all of them together isn't enough when they act in isolation. They have to intertwine, however hard that might seem. The technical wizards need to understand that science is only one angle on life – useful up to a point, but hopelessly reductionist and inadequate on its own. Science offers a complete explanation within its own terms of reference, but thankfully existence is a billion times richer than science knows how to even hint at. Psychologists can help us to engage with the problems, but can't tell us what practical actions might help. Meanwhile,

arts, philosophies and spiritualities alone can't feed the world, preserve the biosphere or hold back a pandemic. Joining perspectives is about more than combining the academic and practical disciplines. It is about linking all the realms of existence, however we might perceive them to be. It respects scientific thinking but also challenges its absolute dominance over the inner voice that sometimes *defies* rational analysis.[2] Tricky but essential.

So, this is my simple mapping out of the skills I think we most need to develop if we are going to thrive from here onwards. It's just a little model. It's hard for me to read over the list because, of course, I'm hit by how bad I am at some of the skills I'm prescribing. Plenty to work on. If you like this list,

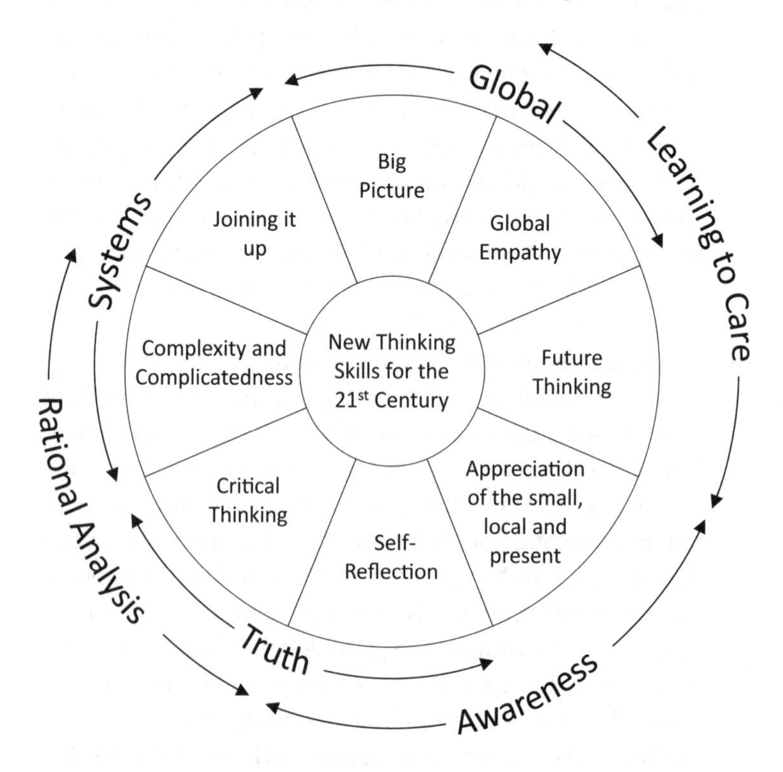

Figure 9.1 New thinking skills for the twenty-first century.

it can also be used, of course, to assess schools, politicians and places of work. Perhaps it is not the entire assessment criteria, but it is a big chunk.

How can twenty-first century thinking skills be developed?

As with all skills, it largely comes down to practice. **Big-picture** thinking can be developed simply by spending time standing right back and pretending you are viewing from Mars. From there, what can we see about how things fit together and what they mean for the world? **Global empathy** involves reading, watching and perhaps even experiencing what life is like for people on the other side of the planet who come from very different cultures. Mix with people from elsewhere if you can. Despite the environmental burden, perhaps even take a trip, but if you do, there is no point in staying in the place that is most similar to life in your own country and sanitizing your experience. If you are going to see and experience it, you need the full immersion. **Future thinking** means contemplating the next generations. **Appreciation** of small simple things and **self-reflection** are both incredibly simple and difficult to do. It is just about stopping and taking the time to be more aware. I recoil from what has now become jargon, but mindfulness seems like a good term for it (I'm hopeless by the way). **Critical thinking** starts with asking smarter questions about your information sources. On page 203 I've outlined some very simple guidelines which might be basic, but if everyone applied them before choosing their news source, the world would very quickly become a better place. Dealing with **complexity and complicatedness** is just a good old-fashioned, head-hurting brain game, that we have always been getting better at and now we need to get even better at it still. Resist the common temptation to shy away from the stuff that matters just because it isn't simple. **Joining up perspectives** is to do with stretching ourselves out of our comfort zones. Do not feel pressure to be good in every area, it is enough that you are taking them on. If you are a scientist or work in a tech giant, make sure you spend at least some time outside of that thought bubble. The first step might even be to notice that there *is* a thought bubble (it is obvious to outsiders). Get into the worlds of stories, emotions, humanities, and other risky scary things that can't be perfectly controlled by humans. Leave the techno-sphere from time to time. On the other hand, if you are an artist, a journalist or a politician get yourself a good-enough scientific understanding of the practical technical and scientific challenges and opportunities that are going on. Get out of your silo, whatever it is. If you are an academic, make sure you are using language that other people can understand so that they can better dip their toes into your world too. Whoever you are, step beyond your specialism and help others to do likewise.

To summarise, to develop these skills we need to get practising. Gulp. Pick a skill. Take 10 minutes to think, talk, read or write about it. Right now? Up to you, of course.

Where is religion and spirituality in all this?

As with the values question, I approach this in a pragmatic way. Do or can spiritual belief systems, communities and practices enable us to thrive in the twenty-first century? Are we at sea without them? I'm not going to try to talk about what might be true and what might not be. I've already made clear that I treat the world as a complex place compared to our finite capacity to understand it, and that we should therefore be cautious in our judgement of different ways of understanding life. We should be tolerant of everything other than intolerance, of course. Beyond that it makes sense that we should be curious and respectful of other ways of seeing things, staying open to the idea that they may enable ways of seeing things that our own current framework keeps us blind to.

In my encounters with the technosphere in Silicon Valley and elsewhere I have been told many times that religion is a big reason why the world isn't dealing with climate change. Fundamentalist mumbo jumbo, I have heard, is getting in the way of the evidence. I can see their point, in that it is frustrating for a scientist to be told that since the world is less than 6,000 years old, anthropogenic climate change can't be real. And there are clearly examples of religions getting twisted to neoliberal, individualist, downright greedy and even terrorist ends that seem wholly at odds with their original principles. Meanwhile, I also see some of the very best thought and action on climate change, for example, coming out of a range of faith and spirituality-based communities. We have also seen the inadequacy of the scientific lens when used on its own, to deal with 'bigger-than-self' problems.[3]

The Buddhist focus on interdependence and compassion, for example, has to be helpful in dealing with global problems. Their treatment of interdependence, which goes way beyond the physical realm and into all aspects of existence, is an alternative language for the joining up of the disciplines that I have been illustrating the need for throughout this book. And in terms of practice, mindfulness surely counts as good training for the development of many of the twenty-first century thinking skills that I'm advocating.[4]

I've singled out Buddhism purely by way of example, because while I'm not a practitioner, its philosophy is one of those that resonates for me. But it is not my intention to privilege it above other forms of spiritual practice. Faith and spirituality-based communities and leaders all over the world are pushing for ways of living that will enable life on Earth to thrive, and helping their communities to gain the kind of big and meta perspectives that we all need to get better at.

Spirituality has gone through such an unfashionable era that I utter the word with trepidation. But the reductionist, techno-science-dominated trend of the past few decades has to be re-evaluated now. It is yesterday's out-of-date thinking. It is no longer clever to scoff at anything that can't be scientifically proven. We already know that science alone, while an internally consistent logical framework, takes us to the wrong place. It can explain behaviour but not sentience. It gives us an incomplete and inadequate framework for living. This much is clear. Enough said, by me at least. I'm not qualified to comment further on this hugely important theme.

10 PROTEST

Do we need protest?

After decades of asking politely and getting nowhere, we have a full-scale emergency on our hands. We have to have change. And it must be now. If the right kind of protest is what it takes, then that's what we must have.

I do not write this as someone who feels instinctive joy at the thought of taking to the streets, but these are serious times. There is compelling evidence that the right kind of protest clearly works. When, in 2019, the UK tightened its carbon targets to 'net zero by 2050', it wasn't far enough, but it was a big step in the right direction. And it looks pretty clear that the political space to make that possible was opened up in no small part by protesters; by Greta Thunberg, by armies of school kids, and by Extinction Rebellion (XR). My work with tech giants, investment bankers, energy companies, an airline and many other corporations tells me that these straight-talking, non-violent direct actions made possible conversations in boardrooms that seemed unthinkable just 18 months before.

What we need is a very special kind of protest. Around the world, but especially in the UK, I think XR has been an incredibly impressive movement. At their best I think they have been brilliant; perhaps one of the greatest causes for optimism in the world today.

They haven't been perfect – but we shouldn't expect that of anyone. They have made mistakes, paid for them and usually learned fast. Anyone who criticises XR needs to first explain what is the more effective game plan that they are already implementing to bring about the system change that we need. XR knows it doesn't have all the answers, but, to quote the partisans in Primo Levi's novel *If Not Now, When?*, it is a case of 'If not this way how, and if not now when?' We owe them a debt of gratitude.

What has been Extinction Rebellion's magic?

It boils down to values, processes, role modelling, positivity and fast learning.

Most importantly for me, XR has been a loud advocate of the three core values I call for in this book. In the London protests of April 2019, you could hear the tannoys reminding supporters to respect everyone: the authorities; the government; the public; each other; the oil companies; EVERYONE. In surreal scenes on Waterloo Bridge, the police were caught visibly off balance by the tide chanting: 'To the police, we love you. We're doing this for your children', as the police carried people away to their vans. Of course, as XR's name suggests, they stand overtly for the preservation of all species. And they made a very strong call for truth.

At its best, XR gave us a taste of a better world. They carried trees onto Waterloo Bridge. They offered free food, interesting talks, a book library and even a skateboard park. There was music and fun. They made the air cleaner. They were nice to everyone – even the few who were less polite in return. They picked up not just their own litter but every piece of litter they could find nearby. No alcohol nor drugs were allowed near the protests. They operate a transparent, inclusive, democratic process, holding debates and votes about every decision, and modelling the evolution in the democratic process that they call for. They were nice to everyone, so that the air wasn't just cleaner from the reduced traffic, it felt kinder too. When they caused disruption they did so respectfully and apologetically. 'I'm really sorry to inconvenience anyone but I'm afraid we do have an environmental emergency and we simply have to have big change right now.' And they were prepared to get arrested.

When they made mistakes, they usually learned from them fast. What were those mistakes? Well to my mind they included:

- Singling out specific politicians (turning up at Jeremy Corbyn's house);
- Graffitiing a Shell Oil building (to me this is at odds with the spirit of non-violence, even though not directed against people);
- Jumping on a couple of tube trains (disrupting the green infrastructure was picking the wrong battle);
- There were very occasional instances of spokespeople straying beyond the scientific evidence, and this undermined the principle of respecting scientific truth;
- Calling for zero carbon by 2025 without any idea of how that might be achieved reduced their credibility – but this is a matter of detail.

For perspective, this is a pretty short list of errors for a large and inclusive organisation.

What is the next evolution of protest?

It is with some humility that I chip in my thoughts as to how XR's activities might be more effective. But for what it's worth here is what I think:

(1) Widen the call for truth beyond just scientific truth about the environmental emergency, to a call for a raising of the bar on truthfulness in public life.
(2) Make sure that the balance of activities is more positive than disruptive. Offer a taste of a better world that does not need to be imagined. Examples might include providing delicious, free, sustainable street food in deprived communities, and cleaning up districts. By doing this XR makes the point that it cares about everyone and stands for a better world. And when it does cause disruption,

people will understand that they are a positive movement, not a destructive force.

(3) Broaden their activities – engaging respectfully with all walks of society, including businesses of every kind. Do this with respect for the people who run such businesses, however unhelpful their activities might be, and even if they are behaving dishonestly. Like Mandela and Gandhi, the trick is to find genuine respect for everyone.

(4) Find ways of lowering the barriers of entry, so that it is easier and less embarrassing for anyone to start to support XR. This might include advice on how to support XR as you shop, or even as you drive.

(5) Continue pushing the deliberative democratic processes and experiments, such as the People's Assembly on Climate Change.

(6) Be very honest and public about mistakes made and the learning process that follows. In this way model what to do when you make mistakes. This especially matters because nearly all our politicians and business leaders have made huge mistakes with respect to the environment and need to be allowed to learn and change.

Should children protest?

The younger generation have been better than their parents at seeing and saying it like it is. If we want psychologically healthy kids we need to let them respond to the challenges of the world as they see them. Striking school kids are playing a huge role.

The kids are right that it shouldn't have been left to them to call out the climate emergency. But my generation hasn't been up to the mark and we should be humbled and grateful for their insistence that we shape up. Why do they so often see and say it more clearly than their parents and teachers? I think it is only partly because it is their future even more than ours. I think the other part of the story is that they have not lived through so many decades of confusing psychological dissonance. All my life the science has been telling us to look after the planet better and society has been pushing me in the other direction. Our whole society has been living in a way that is fundamentally unfit for the Anthropocene, and there has been a collective barrage of fog protecting us from seeing the disconnect between how we are living and how we need to be living. In the words of J. R. R. Tolkien in his classic trilogy *Lord of the Rings*, like Gollum, 'We forgot the taste of bread . . . the sound of the trees . . . the softness of the wind'. But often our kids can still see the disconnect. They see the situation with fresh eyes, and if the emperor isn't wearing any clothes they often find it easier to say so. Greta Thunberg has many amazing qualities, and her youth is just one that makes her arguably the world's clearest voice on the climate emergency.

11 BIG-PICTURE SUMMARY

Rising human power has taken us into the Anthropocene

This is a recent and huge change in the context in which we live. It demands a re-evaluation of how we operate. The Earth is no longer robust to our activities. Compared to us, the rest of the ecosystem gets more fragile by the day. Humans must learn how *not* to expand for the foreseeable future. We will *not* be doing significant space travel for a very long time so we have to make the most of Planet A – which luckily is still wonderful.

We have the opportunity to live better than ever

But the way things are going, a stack of environmental crises threaten to derail us very badly and perhaps very soon. Some of these threats we understand better than others.

One of the better understood Anthropocene challenges that we face is climate change. Much of the weight of effort has gone into the drive for technological solutions. It isn't that we don't need the right technologies, but on their own they won't help us – not even a bit.

The low carbon technologies we need are coming along nicely but on their own they won't help

Food, land and sea present a whole range of other looming crises to get on top of. Technology alone will not solve these either. The biggest levers for change are societal, including diets, population, equality, waste, and how we think about and how we work with the land and sea. We must not use biofuel.

Anthropocene challenges are global, systemic and inescapably intertwined

So far, for all the talk, humans have had zero agency over our greenhouse gas emissions. This does not mean that we *can't* take agency, but we need to recognise that we are not yet pressing the right buttons. Most policy making doesn't take proper account of how the global-system-level dynamics are working and the result is inadequate thinking about what is required – or ignoring the problem altogether.

We need to stand further back from the problem and this entails slowing down more of the time

Habitual ways that humans have used to think and make decisions have proved inadequate to the challenge, and the evidence is that more of the same won't work either. All of this perspective-taking and rewiring requires us to somehow slow down much more of the time. We need to spend more time working up visions of futures that we'd want and which are realistic enough to be exciting.

We need a new system of economics fit for the twenty-first century

We need to think through from scratch the roles of growth, jobs, investments, technology, the way wealth is distributed, the metrics we use and the role of markets. Within this there are big and challenging messages and opportunities for business, including some reframing of the concept itself.

Some types of growth are still healthy but others are not

We have to shrink our environmental impacts. GDP growth is now harmful as a measure of success. Things we do need to grow as fast as we can include global empathy, stewardship, diversity, and quality of life for all species and our capacity for the types of thinking that will allow us to steer our way through the Anthropocene.

We will require globally shared values of respect for all people, for the planet, and for truth

Cultural values and economic frameworks reinforce each other. Values can be deliberately cultivated.

We humans urgently need to develop our thinking skills and habits in at least eight respects

These include big-picture thinking, joined-up thinking, future thinking, critical thinking, dedication to truth, self-awareness, global empathy and a better appreciation of the small things in this beautiful world that we live in.

12 WHAT CAN *I* DO?

When the challenges are so global, and each one of us so small, it can be tempting, but wrong, to think that there is nothing an individual can do to help humans to get a grip. To do so is a cop out. It is one form of human denial of the Anthropocene challenge. The global and systemic nature of our situation does have huge implications for the roles of individuals, organisations and even states. It is true that systemic adjustments can completely undo the direct benefits of many piecemeal actions. So, we need to see everything we do as part of a bigger game.

This means that the individual needs to think differently about their role. What we are looking for is systemic change. Each of us needs to ask the following question:

How can I help to create the conditions under which the world that I want to see becomes possible?

This is a deeper and more thoughtful question than 'How can I live sustainably?', although that is still included within it.

The whole book is laced with more specific ideas, but here are some summary bullets.

- Spend time envisioning the world you want to see. Share that vision and live for it.
- Develop the eight twenty-first century thinking habits and skills as best you can. Encourage our kids to do the same.
- Be as discerning and critical as you can when choosing who and what to believe. Insist on truth everywhere. Reject

politicians, businesses and media sources that don't do likewise. Let them and others know that this is what you are doing.

- Support politicians who demonstrate these habits and skills. Reject those who don't. Be discerning about the difference.
- Exert your influence everywhere that you can: at the ballot box, the workplace, and perhaps most challenging of all, without alienating your friends and family, in every social situation. Remember that since most of us like to be like other people most of the time, by standing up for a better world you make it easier for everyone else to do likewise.
- Remember that every time you invest money you are supporting one future or another. Choose fossil fuel free and otherwise positive banks and pension schemes.
- Be a role model for sustainable consumption as best you can – and find ways of doing so that make your life better. Here is a very short summary of four major ways in which most people can cut their carbon footprint:
  (1) Drive and fly less.
  (2) Reduce net home energy consumption, including by switching from fossil fuels, and by installing insulation, heat pumps and solar panels if you can.
  (3) Reduce meat and dairy consumption, as well as waste. I don't think we all need to go vegan, nor even do it all in one go, but an 80–90% reduction in meat and dairy consumption from 2020 levels is probably about where we need to get to, and if we can do so by 2030 it would be wonderful.
  (4) Find ways to consume less but appreciate more. Buy less stuff and make things last longer. Buy and sell second hand and get things repaired. Borrow and share more. Get to know your supply chains and try to push all your spending power towards organisations that are pushing for a better world.

(The 2020 revised edition of my first book, *How Bad Are Bananas? The Carbon Footprint of Everything*, contains more detailed and recently expanded guidance on cutting your carbon footprint.)

- Don't beat yourself up over your shortcomings but don't let yourself off the hook either. Don't be put off doing what you can by the sense that you might be a hypocrite. It is far better to set challenging standards and fall short than to give up before you start.
- Finally, given the nature of the crisis, even if you don't think of yourself as a born protester, we all have to be asking ourselves very carefully whether or not now might be the time to start, and if so, asking equally carefully exactly how best to go about it for maximum positive effect. It's a very personal decision, of course. (For me, the answer is a pretty clear yes, provided it is the right kind of protest, with the qualities that I outline in Chapter 10 and operating from the values that I call for in Chapter 8.)

What questions were missing? What answers were wrong?

Be part of the process. Add missing questions and answers at www.TheresNoPlanetB.net.

I have set out to ask and answer all the most important and useful questions for life to thrive in the Anthropocene, and to help anyone reading to think through what they might do. Even with a great deal of help, clearly this was never going to be possible. What were the biggest omissions? And what bits do you think I (we) got wrong?

This project has never been a solo undertaking. It has never been about me trying to work it all out in isolation. What is required, of course, is a big collaborative process.

The updated edition of this book has benefitted from about 700 emails of suggestions, and it is with a pang of guilt that I have to apologise for not having replied to them all. Please be part of the collaborative process by chipping in your questions, answers and other contributions. Thanks for your help.

Website: www.TheresNoPlanetB.net

Email: Mike@TheresNoPlanetB.net

Appendix: Climate Emergency Basics

Here are 14 points that I think, for example, every politician needs to understand before they are fit for office.

You could skip this section if you are sure you already know the basics. But just before doing so, perhaps quickly double check that you have fully got your head around everything on my list because even among climate policy makers it would put you in a small minority.

Some of what follows was laid out in more detail by Duncan Clark and I in The Burning Question[1] *although the position briefly outlined here also contains some important updates.*

STOP PRESS As this book went into its final edit, the IPCC's long awaited special report 'Global Warming of 1.5 degrees' hit the front pages of all good news media.[2] It is the IPCC's most urgent call for action so far and is a very useful development. Since it draws mainly on the same source material that I have used, it is no surprise that the report is totally coherent with the points in this Appendix. Happily, unlike the IPCC, I have not had to negotiate the content with any politicians, so it is easier for me to get quickly to the point without any requirement for tact.

Point 1: A global temperature rise of 2 °C looks very risky but 1.5 °C much less so

In truth, no one really knows how bad the consequences of any particular temperature rise might be. We don't have a good understanding of the various potential tipping points that we might trigger in the environmental system, nor how successful humans might be at dealing with them. The uncomfortable truth is that when we meddle with the climate, we play with stuff that we don't really understand and which can't be put straight if we mess up. Even 1.5 °C might be enough to trigger some dramatic change in the climatic conditions, such as an unstoppable flow of methane boiling out of melting permafrost or a collapse of the ocean ecosystem. On the other hand, it is also just about conceivable, although very unlikely, that 3 °C might not be too bad to live with. Most climate scientists are confident that 4 °C would have very nasty consequences for humanity.

There is widespread scientific agreement, endorsed by just about every country (including the pre-Trump USA), that a temperature rise of 2 °C would be very dangerous and to keep to a comfortable level of risk, we should cut emissions so as to limit temperature change to just 1.5 °C. Much climate modelling assumes that the temperature change is roughly in proportion to emissions. But it is not at all clear that this will be the case. It is also very possible, perhaps even likely, that at some point we trigger 'positive feedback mechanisms'; vicious circles in which temperature change causes things to happen that in turn trigger more temperature change. This would be likely to provoke a step change in the climate that would probably be unstoppable by human activity. A recent and credible paper looked at five of these positive feedbacks and estimated that the trigger point for a step change could well occur at around 2 °C.[3]

As I write this, at +1.1 °C and rising, we are seeing unprecedented climate change induced wildfires in Australia adding around 1% to global emissions, and thousands of craters up to 50 metres across where methane has exploded out of the permafrost.

Point 2: As long as we don't trigger a step change in the climate, temperature rise corresponds roughly with the total amount of carbon we have ever burned

In other words, it is the cumulative emissions that matter most. Any amount of temperature rise will be determined, roughly speaking, by an all-time *carbon budget*. Nobody knows exactly how much carbon results in an exact temperature change, and therefore exactly what those budgets are, but thanks to some highly sophisticated climate modelling, we have quite good ball-park estimates. In fact, the link between carbon emissions and climate change turns out to be at least as predictable as many of the standard economic forecasts that politicians take very seriously, such as GDP growth or unemployment rates.

This cumulative budget method of thinking is very useful, but is only an approximation. Other greenhouse gases, such as methane, also have an impact and very importantly affect the speed with which the global temperature rises. All the carbon budget estimates relate only to carbon dioxide (CO_2) and are based on background assumptions about what will be happening with the other greenhouse gases.

Point 3: Emissions of carbon dioxide, the most important greenhouse gas, have grown exponentially for 160 years

When I say 'exponential', I don't just mean any old banana-shaped graph. The carbon curve has always had an uncannily steady long-term annual growth rate of 1.8%, which means emissions have been doubling every 39 years. It really is exponential in the mathematical sense of the word. Why this is so interesting is because a property of exponential curves is that when they double, they do so in many different ways at the same time. In particular, when the height doubles (in this case representing annual carbon emissions), the steepness also

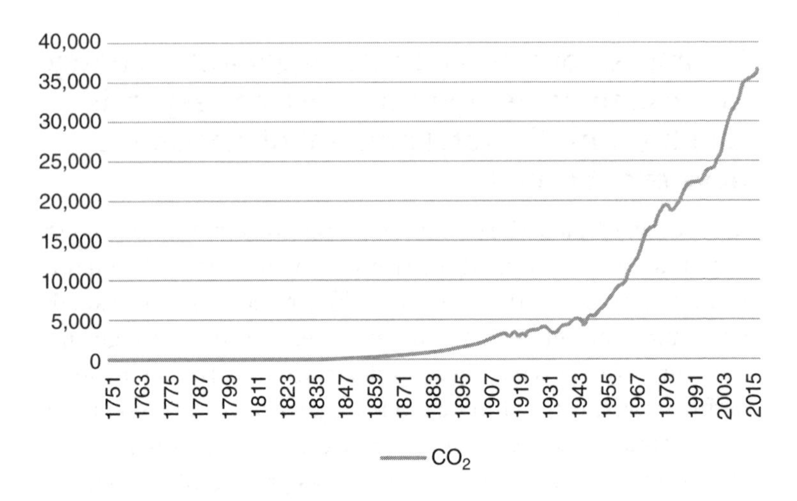

Figure A.1 Annual global CO_2 emissions 1751–2018.[4] CO_2 emissions have increased sharply since the beginning of the twentieth century.

doubles (in this case representing the annual growth in carbon emissions) and the area underneath the curve doubles (representing the sum total of all previous carbon emissions – known as the cumulative emissions). So, for our carbon curve, annual emissions, annual growth in emissions and the sum of all the emissions we have ever produced have all been doubling, more or less like clockwork, every 39 years.

(Of course, the curve isn't totally smooth, it has a small amount of noise. There are ups and downs between individual years, and even sometimes periods of a few years at a time that are below the trend line, but they are preceded by and followed by a few years above the trend. See the endnote after this sentence for more about this wiggling which, while very interesting, is a distraction from the key points.[5])

Point 4: We have not yet dented that carbon curve

Some people got quite excited about three recent years of flatlining emissions between 2014 and 2016. Sadly, there was little statistical significance in this.[6] In other words, the slight dip below the 1.8% growth trend line was well within the usual noise. Remember that over the years there have been plenty of times when the carbon has dipped for a while before flying up above the trend line again. So, these flat years were entirely consistent with business as usual, even though, looking on the bright side, it was also true to say that they looked consistent with us having taken the first steps towards taming carbon emissions. They didn't tell us much either way.

Then 2017's emissions jumped 2% higher, putting to bed any grounds for optimistic over-interpretations of the previous three years. In 2018 emissions rose by 1.1% and in 2019 by 'just' 0.6%. This is not strong evidence of a trend. There is still little or no basis for the claim that humans have yet dented the curve. Actions to date have resulted in zero or almost zero agency. Stark but true. And if we want the situation to be otherwise, we'd better face it.

Point 5: At the current rate of carbon emissions the remaining viable carbon budget for both 1.5 °C and 2 °C is dwindling quickly – despite some recent good news from the carbon modellers

In a rare piece of good news from the climate science community, in 2017 a re-running of the climate models using more up-to-date data estimated a little bit more carbon in the budget for both 1.5 °C and 2 °C than had been thought.[7] Don't relax though, because all this means is that it now looks as if rather than being on track to overshoot our budget for 1.5 °C in 2022 (Gulp!), at today's emissions levels, we might still have a 66% chance of not doing so until somewhere between 2030 and 2040. Some of my colleagues think this is a trivial difference, but I think we should take good news wherever we can find it. It is significant enough to make the 1.5 °C ambition look a little less far out of reach. Just to be clear, there is no room whatsoever for complacency.

(Very roughly, 2.2 trillion tonnes of CO_2 is now thought to raise the global temperature by about 1 °C, whereas we used to think it was more like 1.8 trillion tonnes per degree.[8])

On the more sobering side, a continuation of the traditional doubling of cumulative carbon every 39 years would mean that 4 °C would be only 39 years behind the 2 °C threshold and 8 °C only a further 39 years behind that.

Point 6: It takes a long time to put the brakes on

Steering the climate is not like steering a racing car. It is no good waiting until we start getting nasty weather before taking action, because our climate reacts like an oil tanker. To get the point, imagine if everyone in the world decided right now that we had to put the brakes on climate change as fast as we could. We would first need to make a plan and then start implementing it before emissions would even start to fall. During these phases, we'd still be making things worse, and every day committing ourselves to yet more temperature change. Even once we reached zero emissions, we would have to wait further, with our fingers crossed as to the severity of the symptoms we encountered, while the temperature carried on rising for a while. Because the world's ice will continue melting long after the temperature begins its long journey back towards preindustrial levels. *If* we succeed in developing the technology to take carbon out of the atmosphere, we might then stabilise temperatures quite quickly and even hope to nudge them down again, but without this, we would still have a few more decades of further rise before equilibrium were reached.

The captain of an oil tanker thinks a long way ahead when considering a change of course, but as a species, we are not yet good enough at doing that. This is one of the most serious skill deficits for humanity in the Anthropocene: the poor capacity for planning ahead.

Point 7: All the fuel that gets dug up gets burned, so it has to stay in the ground instead

Amazing that such an obvious point has taken the political world so long to start getting its head around – and it isn't there yet. Once fuel leaves the ground, it all gets burned to meet a consumer need. The carbon footprint of extracted fuel is just about equal to the carbon footprint of burned fuels and the carbon footprint of all consumer goods and services. It works like three carriages of a train coupled together. They push and pull each other along, ensuring that they all travel at the same speed. Either we slow them all down or nothing will happen.

The only real exception to this is a trivial proportion of extracted oil which goes into making plastic, however, adding to the already vast amount that is clogging up our planet (see the discussion about plastics on pages 63–65).

As I update this book in March 2020, the UK is bewilderingly on the brink of committing to a new coal mine, and in doing so looking more than a little foolish in the run-up to the COP26 climate talks as it tries to claim global leadership on the climate emergency.[9]

Point 8: Many of the things we might assume will help haven't

The 'balloon squeezing' or, to give it its proper name, 'rebound' effect describes the unfortunate tendency of savings in one place, only to get counteracted by adjustments elsewhere in the system. A great many people who think they understand about rebound effects still vastly underestimate their true significance. The reason for this is that you can't fully quantify rebound effects by adding them up one by one as they work by sending ripples through the whole economy. There is actually an infinite number of potential rebound pathways. As an example, just to give a sense of how it works, if you buy a more efficient car, here are just some of the ways in which the carbon

Figure A.2 The balloon squeezing effect: when emissions are squeezed in part of the world or part of the economy, the emissions from the rest of the system expand to compensate. So if we want change, we need a way of squeezing everything at once.

savings might get lost; you might drive further, you probably spend any money you save on other things that have a carbon footprint, the fuel stations adjust their prices slightly and sell more to others; car manufacturers adjust their marketing pitch to sell their higher carbon cars to others, the oil industry adjusts its sales and marketing pitch towards other people and other countries, you become more likely to live further from the city in a larger property that requires more heating, your increased mileage increases the requirement for road maintenance . . . and so on. You can't possibly list every one of these effects, let alone quantify each one. Attempts to do so are doomed to underestimate the total rebound phenomenon.[10] However, we can perhaps do better if we stand right back and look at things at the global system level. In fact, the resolutely exponential nature of both the carbon and the energy curves (1.8% and 2.4% growth per year, respectively) mean that we can say that overall carbon and energy rebound effects of all the world's efficiency gains combined have been 101.8% and 102.4%.[11] Now it is suddenly clear why it might be that we use more energy than ever before, not just *despite* all the efficiency improvements we have had over the years but perhaps even *because* of them.

Just before you go ripping out all your double glazing and deflating your tyres, note that none of this means that efficiency gains cannot be useful in the future. But it does tell us that efficiency needs to be coupled with a constraint on total resource use. Unless we bank the savings they make into productivity increases.

Point 9: The world needs to use less energy

Even growing our renewables to the size of today's global energy supply will do us no good whatsoever if we stay on today's energy growth trajectory. To give renewables a chance of replacing rather than augmenting fossil fuel, the world needs to use less energy.

We humans have always wanted more energy than we have got. Since the pyramids were built with human slave power, we have always been hungry for more energy. New technologies and efficiency gains have enabled us to increase the supply more or less continuously. If this process continues, we might still expect an explosion in renewables to bring about a significant temporary dent in our appetite for more energy, and to reduce our hunger for fossil fuels somewhat during a period in which we might feel relatively awash with energy. But this will not be anything like enough to keep the fuel in the ground. Renewables can make it easier for us to stop using coal, oil and gas, but they will not make it happen on their own.

(In Tibetan Buddhist mythology, to be a *hungry ghost* is worse than being in Hell. It is a state in which the more you are fed, the hungrier you get. Hell is not quite as bad. What might happen to you there, for example, is that molten copper might be poured down your throat after which you will be magically breathed back to life so that the process can be repeated. It's grim, but temporary. As a hungry ghost, however, you have no respite from the torture of your own appetite and the downward cycle is almost impossible to break; the more you get what you want, the more you suffer. This is us. Energy addicts. This is why one of the central questions of this book is how we can achieve satiation.)

Point 10: We urgently need a working global agreement to leave the fuel in the ground

Piecemeal actions by individuals, companies and countries won't cut the carbon emissions on their own because of rebound effects. The easiest place to put the brakes on is at the point of extraction. However challenging this might be it is still the easiest way. Constraining carbon at the point of emissions is also a possible route, but it is harder to monitor, and provides a lot more possibilities for the balloon to expand in the places that we fail to squeeze.

Point 11: We need to manage other gases too

As if the carbon challenge wasn't tough enough on its own, the other greenhouse gases are important enough that we won't solve climate change unless we deal with them too. In particular we need to deal with methane and nitrogen dioxide. This means we need to look at what we eat and how we farm it. Less animal production, especially less ruminant animals (e.g. cows and sheep). More judicious use of fertiliser. Cut emissions from landfill through less landfill and better landfill sites. (See Chapter 1, Food.)

To give an idea of how important other gases are, the latest modelling results suggest that the difference between strong action on non-CO_2 gases and less strong action would be to add about 40 Gt of carbon to the remaining budget for 1.5 °C. That is around four years' worth of today's CO_2 emissions.[12] Remember that all the carbon budgets are based on assumptions about levels of action on the other gases as well.

In numerical terms you could think of non-carbon emissions as being something like one quarter of the climate change problem, but in a way that is misleading because we can't deal with the climate emergency without sorting them out as well. They are the poor forgotten relations to carbon, but too influential to be ignored.

In fact the story with the other greenhouse gases is a bit more complicated because they act differently to CO_2. CO_2 stays in the atmosphere, warming the planet for many hundreds of years. Methane has a much more powerful effect but a much shorter life. If you are interested in the temperature change in 100 years' time, 1 kg of methane has the same impact as about 25 kg of CO_2, and this has led to a widely adopted convention that 1 kg of methane can be considered equivalent to 25 kg of CO_2. However, if, as might well be the case, you are interested in the temperature we might reach just 50 years from now, then you have to consider methane to be twice as powerful a greenhouse gas as we usually think of it as. Then again, if you only care about the final temperature we might end up at in a few hundred years' time, then you could more or less discount the methane entirely.

Point 12: Extracting and burning fossil fuel has to become too expensive, illegal or both

The easiest way of making it too expensive is to have a carbon price (and it will need to be a big one; a few hundred dollars per tonne quite soon). People who say this is too difficult to achieve fail to put forward equally effective solutions that would be effective enough. At least with a global carbon price, all the difficulties are up front. Once it is in place and backed by adequate enforcement mechanisms, the hardest part might be over.

(See the discussion about the carbon price, page 166.)

Point 13: The global deal will need to work for everyone

Everyone has to be on one side, because just a few weak links will break the whole thing down. This makes it an incredibly challenging deal to bring about. But the difficulty does not make it any less essential.

At the moment it is very tempting for politicians in many countries to take the view that it is not in their countries' interests for the world to get on top of the climate emergency. Some countries will have to let go of assets while others will find that their abundance of renewable energy resources will give them a head advantage in the low carbon world. Meanwhile, the threats from climate change will hit some hard early on while others experience benefits in the short term. While the Maldives sinks and Bangladesh floods, Russia is likely to find its crop yields going up at first, its ports becoming ice free for 12 months of the year instead of eight, and yet more of its fossil energy reserves becoming accessible. In poorer countries a carbon constraint could impact on wellbeing more seriously than in richer places where the link between energy and happiness has probably already been broken in the same way that the link between wellbeing and GDP has been shown to break down.

A global deal is going to be challenging to reach because it requires both understanding of the different implications for each country and a sense of international fair play that the world has never yet known. The difficulties do not change the reality that we must have that deal.

Point 14: We need to take carbon back out of the atmosphere

Our ability to apply the brakes is so poor, and the need so urgent that is clear we will need a reverse gear. I'm not just talking about taking carbon out of factory chimneys. What

I mean is that we need to develop ways of taking it out of the air that we breathe – direct air capture (DAC). We don't quite know how to do it on the required scale yet, but it looks likely that serious investment would sort that out too. Even with improved technology, it will always be costly to do, but a carbon price can sort out that problem. At the moment, Climeworks has a plant in Iceland which is capturing a tiny 50 tonnes per year, at a cost of around $1,000 per tonne. They think they can slash this to about one fifth of the price as they scale up. Meanwhile, Carbon Engineering in the USA claims that $100 per tonne would be feasible at scale.[13]

Finally, on this point, do not fall into the trap, or allow others to do so, of thinking even for a second that carbon scrubbing might mean we don't need to be so concerned about cutting emissions.

ALPHABETICAL QUICK TOUR

This section gives you a super-speedy dip into much of the ground covered in the book. If you want to read just one paragraph before you go to sleep, this might be the section to open up. I have also used this section as a chance to include a few things that didn't fit anywhere else, but deserved a paragraph somewhere. I have missed a lot out.

By putting things in alphabetical order, everything gets jumbled together in a new way, reemphasising the interconnectedness of everything, and the different ways in which it all fits together.

Aeroplanes

The essential means by which people from different parts of the world get to meet face to face and by which a world population gets the opportunity to experience other cultures and people first hand. Unfortunately, the A380, the world's most efficient passenger aircraft, requires around 450 tonnes of fuel to fly one way from London to Beijing and at the moment there are no foreseeable technologies for replacing the burning of liquid fuels to power aeroplanes – at least for long haul. So, the options are fossil fuel (but see Fossil fuel, below), biofuel (but the trade-off against food is incredibly steep, see page 88), don't fly (but that is very painful – see Hair shirts, below) or create liquid fuels from renewable electricity (which is feasible, but inefficient; see page 81). The pathway through all these Hobson's choices looks like being to fly more frugally, develop fuel from solar power and electrify short flights.

Animal feed

Here we are not talking about grass or pasture. We are talking about plant matter that is digestible by humans but is instead fed to animals. Globally, there is enough of this to provide about 1,800 calories per day to every person on the planet; that's about three quarters of the global food requirement for human energy. In return, animals give us back about 10% in the form of meat and dairy products. Cutting out animal feed would hugely help us feed the world and preserve biodiversity.

Anthropocene

The age in which we have become big people on a small planet rather than the other way round. The age in which human activity is the biggest thing affecting the environment and the climate. We entered it recently and it necessitates a big adjustment in the way in which we carry ourselves as a species – an adjustment that we are so far struggling with. Our blundering arrival in the Anthropocene is the reason for this book.

Balloon squeezing

More properly known as the **rebound effect**, this is the way in which the effect of micro-actions can be nullified by counter-balancing adjustments elsewhere in the global system; the rest of the balloon expands to compensate for one bit getting squeezed. Balloon squeezing accounts for the brutal way in which the sum of all local and national carbon savings to date have led to either no or at best only very slight change in the global emissions trend. It is not possible to list one by one all the different ways in which rebounds occur and consequently many people play down the overall effect. All the serious analyses of this that I have seen that say energy and carbon efficiency rebounds are less than 100% fail to take account the macro-system-level effects.[1]

Balloon squeezing happens at the national level when cuts in emissions are accompanied by equal and opposite increases in other countries. Prominent examples from the past few years include the UK getting more of its products made in China, and the USA exporting more of its coal for others to burn.

Biodiversity

The variability of living organisms, applied to either a particular place or globally. Essential for healthy life on Earth, and currently crashing through the floor due to a variety of (mainly human) impacts, for example pollution, habitat destruction and climate change. In the UK, farmland bird populations decreased by 56% between 1970 and 2015. Flying insects in German nature reserves declined by 76% over 27 years (1989–2016).[2] Biodiversity is very hard to replace once gone. (See page 60.)

Biofuels

Energy from plant matter. There is a limited place for this in a sustainable energy system; managed woodland for burning fuel, conversion of waste products to liquid fuels, and perhaps a small amount of cropland for biofuel to allow aviation in the low carbon world. If land is used for biofuel crops instead of food crops, the nutritional sacrifice is huge for just a small contribution to the world's energy supply. An explosion in biofuels as a replacement for oil would present a real threat to global nutrition. (See Are biofuels bonkers?, page 88.)

Bullshit

Any attempt to nudge others towards an inaccurate understanding of what is going on. Beyond lies and fake news this includes all deliberate misrepresentations of fact analysis. Famous examples include attempts to confuse the evidence that climate change is a big deal and the suggestion that Brexit would liberate £350 million per week for the National Health Service.

Here in the Anthropocene we can't afford to make things any harder for ourselves than they already are, and we need a major purge on bullshit. All of us need to get better at differentiating fact from fiction, smarter at working out who to trust, and less tolerant of those who we cannot trust. It applies especially to media, politicians, business leaders and colleagues.

The Burning Question

Duncan Clark and I wrote a book with this title as an introduction to climate change from a global system perspective. If you want details on exponential energy and emissions curves, rebound effects, why we need to leave the fuel in the ground and what has been stopping us, then here is one place to look. Although written before the Paris climate summit, and we have burned even more fuel in the seven years since its publication, there is barely a word of the analysis that I would want to correct if we were writing it today.

Business

The source of most of our environmental impacts and hugely influential on human desire to live less or more sustainably. Also, the source of a great deal of much self-congratulatory greenwash, which is one form of bullshit (see above). For perspective, the sum total of business response to climate change has so far resulted in a growth in business carbon emissions. So there is no scope whatever for self-congratulation. On every environmental factor, businesses need to do three things, no two of which are enough: (1) cut their impact down, (2) produce goods and services that enable others to do likewise and (3) push, in whatever ways they can, for a global arrangement to cap impacts, especially inputs and outputs to the environment, such as fuel extraction and carbon emissions.

Carbon capture and storage (CCS)

Important technology but not a game changer. Taking carbon from big point sources like power stations is helpful while still using fossil fuel, and applied to biofuel power stations could even lead to negative emissions. Storage is not without risk, but then nothing is. It costs money and therefore requires a carbon price and/or subsidy in order to take off.

Carbon scrubbing (or direct air capture)

Potentially a game changer, with the potential to take carbon out of ambient air, either using land and plants or through 'synthetic trees'. We could really do with it, but at the moment it is still an expensive emerging technology. Will big R&D investment lead to a rapid solution? Will it be like building the bomb or more like curing cancer? We should find out.

China

Around one sixth of the world's population and almost one third of the world's carbon emissions (and rising). Free from many of the strengths and the weakness of today's democratic systems, perhaps it can lead on many of the things that matter most, or perhaps it will be deeply damaging. More than in most countries, if a policy idea is seen as a good thing, the Chinese can bring it about.

Climate change

One clear symptom of humankind's clumsy arrival in the Anthropocene. Not the only one and it won't be the last. In dealing with it as a specific issue, we would be very wise to develop the skills and structures that will help us to deal with future Anthropocene issues, including any that give us less time to react. 'Climate emergency' is often a better expression to use, being just as accurate and more descriptive.

Coal

The cheapest, most polluting and most abundant fossil fuel. The one that most urgently needs to stop coming out of the ground in every single country. This requires a global deal involving everyone. We also need every country to stop burning it. And we need every supply chain to ensure it is coal free. Any climate policy or business strategy that is not consistent with movement in this direction should be thought of as a non-starter.

Commuting

The way we get to and from work can be anything from an avoidable, dull, polluting, expensive waste of time to a joyful, life-extending, sociable and eco-friendly part of life. The difference between the two often boils down to a simple bit of creativity and habit breaking. Think lift-shares, bikes, walks, electric vehicles, trains, buses and home-working.

Consumerism

The name of an era that has to come to an end. A poor but seductive substitute for happiness to which no one is immune. The more any of us get drawn into this, the more we should treat it as a warning sign that something else in your life isn't right.

Cycling

In the absence of solar electric bikes powered entirely by renewables, this is the most efficient form of land transport known to woman and man. Wonderfully healthy, as long as you don't have a crash or breath in too much diesel pollution. Recent UK research has found that the cycle commute cuts the premature death risk by 40%.[3] Amazing. Against that just 100 cyclists per year die on the UK's roads (that is a small number compared to 160,000 from heart and circulatory diseases, approximately 164,000 cancer deaths and 1,775 total road deaths). And the more people who cycle, the safer it is.

The word of caution about pollution is because it causes an estimated 40,000 premature deaths per year, and a cyclist on a busy urban road is both badly exposed and is almost certainly breathing more deeply than most people nearby. Britain's love affair with diesel cars (now firmly over) made our problem quite a lot worse than it needed to be (see Diesel, below).

Interestingly, the solar-powered bike will comfortably beat the conventional bike in terms of the land required to generate the fuel to power the machine (see page 117).

Democracy

The best system of government known to humankind perhaps, although there are people in China who observe and challenge. It relies on thoughtful, accurately informed voters, and even then isn't supposed to be about the public being expected to vote directly on highly detailed and complex policy decisions. Clearly not working well enough at the moment in, for example, the UK or USA to deliver 'Anthropocene-fit' politics. It looks like both politicians and voters need to raise their game and/or we need an evolution of the concept itself.

Fit for purpose democracy entails not just voting, but accurate information, and a widespread sense of responsibility for the common good.

Determinism

Whether or not there is such a thing as free will is not something that can ever be proved or disproved. I go about life as if there is such a thing, and I bet you do too. However, I do encounter a lot of determinist thinking around the Anthropocene. It comes in two forms, and both of them result in disengagement from the challenge. Either 'We will be fine because we always have been', or 'We've had it because humans are too short sighted and selfish to survive the Anthropocene.' The first statement could also have been made by the dinosaurs in their day. The second statement does not prove that we can't change in time. Because it is not proven that we must fail, the fight is worth taking on and any other approach is lazy and cowardly.

Developed countries

These are the places where human impact on the planet per capita is highest. Home to most but not all of the super-rich as well as a few, but not most, of the world's poorest 10%. The birth places of most of the technologies that are tearing the ecosystem apart as well as most of the ones that could help us manage it properly. The parts of the world that most urgently need to start discovering how enough can be enough, but not necessarily the home of the best thinking on those lines.

Diets

For most people reading this, a sustainable healthy diet means: (1) less meat and dairy; (2) especially less produce from cows and sheep (the methane-burping ruminants); (3) more fruit and vegetables (keeping them local-ish or shipped); (4) less sugar and salt; and (5) plenty of variety.

Double-sided photocopying
An example of, and metaphor for, very small resource efficiency measures that should, of course, be routinely adopted, but which are so small as to undermine, if mentioned, the credibility of any corporate responsibility statement.

Economics
A discipline in need of a make-over. Mainstream economics, partly through inadequate metrics, locks us into harmful forms of growth and nudges us into unhelpful values. Difficult to change because it is such an entrenched framework by which so many people, organisations and countries understand the world. (See Chapters 5, 6 and 7.)

Education
The primary means by which we enable the next generations to deal with the issues we have left them with. Anthropocene-ready adults will demonstrate the eight thinking skills listed in Chapter 9.

Efficiency
A quality that humans have developed continuously for millennia and which has always gone hand in hand with increases in total consumption, resource usage and environmental impact. It has always been and still is integral to the dynamics of growth, with all its benefits and, increasingly, its problems. Going forwards if we can put a firm limit on the environmental burden we place on the ecosystem, including capping the carbon, the role of efficiency will change fundamentally, and we will become able to call it, without reservation, a force for good. It could then be the means by which we can do more, if we want to, with less impact.

Electric cars
A must-have new technology but not a silver bullet. They pollute less, they can use somewhat less energy and, more to the point, in

future that energy can come from renewables, even though, for now, more driving still means more fossil fuel use. Like their fossil fuel predecessors, electric cars are hugely resource intensive in both their manufacture and use, both of which need to be done more sparingly.

Emissions

Carbon emissions over the past 160 years have been uncannily exponential with a growth rate of 1.8% per year (see Exponential growth). Some people got excited recently about three years of flatlining, but it was never statistically significant and at the time of writing the latest annual figures are above the trend line at 2%.

Energy growth

The more we have, the more we grow our use. Human energy growth has hovered at around 2.4% per year for the past 50 years. The growth rate has actually increased from 1% in the previous hundred years and lower still before that. So it looks super-exponential. While some people speculate that humans might be reaching the end of their appetite for more, without the need for conscious moderation, this looks like wishful thinking in the extreme to me.

Enoughness

There is a time for everything to stop expanding. Humans have reached it in a few key areas. One of the crunch questions for our species right now is 'How can what we have feel like enough if it becomes unhealthy to want yet more of it?' There is still plenty of room for ambition in life – we just need to direct it in positive ways.

Evolution (of humans)

Something we need to rebalance. We have developed the technology and the power. Now we need to complement that with new sets of thinking skills; global empathy, systems thinking, appreciation of what we have and values that allow life in the global village where all the eggs are in one basket, like it or not.

Experts

As the complexity of the issues goes up, so we need to get better at understanding what experts tell us. Experts need to get smarter in their specialisms but at the same time more joined up with other disciplines and better at communicating to anyone who is interested. It's a challenging consequence of technology.

Exponential growth

A growth rate that is proportional to the amount that already exists. If you put two foxes on an island full of rabbits, the growth rate is exponential as long as there is no shortage of rabbits and nobody else takes action to control the growth. Money in the bank might grow exponentially, and so might your debt – unless someone intervenes to change it. An exponential curve is not just any banana-shaped curve; it is one in which the steepness is mathematically proportional to the height. Whenever you get one, it is evidence of a positive feedback mechanism in which the more there is of something, the faster it grows. (See Emissions.)

Facts

Things that need honouring as never before. It is hard enough to see what is going on without deliberate or careless misrepresentations by media, politicians or business. Every citizen can help, at almost every turn, to insist on the cleanest possible view of the facts. From the office desk, to the voting booth, to the choice of news station to the bar chat.

Fake news

A phenomenon that has always been with us but which has recently erupted with devastating consequences. The tools for its propagation get more sophisticated, all of us have to improve our capacity to spot it and the vigour with which we reject it. Many of the thinking skills I outline for twenty-first century living are to do with raising our defences against fake news.

Farming

Growing food with respect to both world and wildlife has to be one of the most useful ways that a person can spend their days. A free market incentivises the minimisation of the number of people required to produce a given yield, but this is madness, because it takes care, attention, skill and hard work to get this right. And we do not have any shortage of people. In fact, we have more than ever before with at least a couple of billion more on the way. We need more people working the land, and doing it properly.

Fish

Global consumption runs at about 19 kg per capita per year.[4] An important source of protein and micro-nutrients, especially in many poor communities for whom it is essential that their catch stays out of the global food market. While wild fish are a scarce resource, threatened by over-fishing and ocean acidification from the burning of fossil fuels, farmed fish incur all the inefficiencies of feed-fed land animals as well as methane emissions from rotting waste and food on the sea or pond floor. Eat in moderation, if at all, and with good understanding of the supply chain.

Food system

Almost 6,000 calories are grown per person per day and there is an even greater abundance of protein. But 1 billion don't get the nutrients required for health while twice as many eat too many calories. Current food production is enough for 9.7 billion in 2050 but only if we cut the animal feed (reducing meat and dairy), reduce waste and limit biofuels. And then we need to distribute it properly. There is no physical reason why all this can't be done while simultaneously improving biodiversity, especially if more people become farmers.

Fossil fuel

Almost all of this must stay in the ground or be put back there again after burning. This is uncontentious and no politician that fails to be clear about this is worthy of a vote, no matter what they say about anything else. No company executive that fails to grasp it is fit for their job. And no friend, family member or drinking companion who still thinks in terms of a fossil fuel future should go unchallenged. We cannot afford anything other than clarity on this simple point. Along with 'Drive on the correct side of the road', 'Remember to charge your phone' and 'Brush your teeth at night', store 'Keep the fuel in the ground' in the 'boring but important' category of your brain for frequent reference.

Fossil fuel companies

Arguably, we might owe a great debt to fossil fuel companies for enabling today's affluent, techno world. However, the context has now changed and they need to evolve or die at high speed.

Fun

To say that there is no point saving a world that is no fun to live in doesn't really mean that we need to be laughing all the time. But it is my shorthand for saying that life needs to feel worth living – and as much so as possible. If a proposed change feels like putting on a hair shirt, it is probably a sign that it hasn't yet been properly thought out. Overall, one planet living has a lot more to do with opportunity than cost.

Fracking

In principle, this technique for taking natural gas out of shale rocks could have a marginal role to play as a transition fuel on the way to the low carbon world, because gas is a less carbon-intensive fuel than coal or oil. However, the extraction process requires energy and there are serious environmental issues to get right. Without both very tight regulation and a trustworthy assessment of the risks and benefits, it should certainly stay in the ground. The UK, for example, is currently a long way from meeting these criteria.

Gas (natural gas)

The least carbon-intensive fossil fuel in terms of energy per tonne of carbon dioxide. Emits about half the carbon dioxide per unit of energy that oil does, and is even better compared to coal. Sadly gas and oil extraction usually go together. The exception, fracking, is marred by other environmental concerns. (See Fracking.)

Geo engineering

Technical solutions to the climate emergency other than stopping emitting or taking carbon back out of the air. Geo engineering solutions range from the very risky, untried and downright foolish, to the feasible but limited.

The most sensible looks to be solar radiation management using sulphate aerosols. This would involve spraying sulphate into the stratosphere in order to reflect sunlight and offset global warming. The arguments in favour of this include its potential low cost and low-tech approach, as well its mimicry of a natural process (i.e. when sulphur dioxide gas is released during volcanic eruptions). However, the best it can achieve for us is limited – a number of possible side effects include ozone depletion, changes in solar radiation (a potential catastrophe for plants), changes in precipitation, and perhaps some human health effects.

Global dynamics

The way the world economy and our energy use works at the global system level is not discussed enough. When you stand right back, some incredibly important things hit you, with huge consequences for climate policy and, more widely, humankind's entire response to the Anthropocene. Efficiency, for example, leads, more often than not, to an increase in resource use rather than a saving. See also Efficiency, Rebounds and Technology.

Global governance

This is something we need enough of so that we can manage global challenges. This is not instead of regional, national, local, family or personal governance; it is as well as. We need governance at every level at which we have responsibilities and interests. Relatively recently we could get away without including global in the list. Here in the Anthropocene, we can't. That we find it difficult is evidence of the need for more practice.

Greed

A simple term for individualism. And one of the seven deadly sins. Along with hate and delusion, one of the three roots of suffering according to Buddhist philosophy. Greed is the motivation that drives the neoliberal agenda. The opposite of empathy. Would it be simplistic or concise to say that overcoming greed is the secret to wellbeing in the Anthropocene, and everything else in this book is just detail?

Greenwash

Prevalent, sometimes through wishful thinking, and sometimes deliberate. The easiest way for an environmental consultancy to make a living is to collude in this, and even if you don't intend to, it is hard to know when you have been sucked in. We all need to be on the look-out and challenge it when we see it. The key test is to stand back and ask carefully 'Does this take us closer to humankind surviving the Anthropocene?' If not then its greenwash.

Growth (see also Exponential growth and Energy growth)

Some forms of this are more essential than ever before (global empathy, the capacity to appreciate life without high material consumption, critical and systems thinking), while other forms that we have taken to be normal or healthy now look as dangerous as hell. Somewhere in the middle, and highly contested, lies growth in GDP.

Hair shirts

Largely unnecessary in the battle against the climate emergency as well as the wider challenge of humans responding to the Anthropocene. Sadly, it is still what many people think about when they think of looking after the environment and creating a society in which everyone can live well. The trick, in the avoidance of hair shirts, must be to ditch the things in life that do not contribute to wellbeing, to use more imagination and to focus on the wellbeing opportunities that arise from growing beyond the twenty-first century rat race.

How Bad Are Bananas? The Carbon Footprint of Everything

A quick plug for my first book, which is one place to turn if you want more info on carbon footprints. Since publication in 2010 some numbers have adjusted a bit; I'm much more supportive of subsidies for micro-renewables and I would take a stronger line on plastic today, but apart from that the broad principles still stand.

Human denial

Fascinating and hugely important phenomenon by which humans (all of us, you and I included, I'm afraid) become able to demonstrate incredible mental dexterity to avoid confronting facts that give us unpleasant emotions when we think about them. Well understood in the worlds of therapy and grief counselling, denial has been fundamental to our species managing not to deal with climate change despite irrefutable evidence that we need to pull our fingers out.

Denial of our responsibility to take action on the Anthropocene challenge takes many forms including: ignoring it, wishful thinking about possible solutions (such as technology alone, migration to other planets), determinism, small actions to salve the conscience.

Human psychology

An essential component of the big puzzle that doesn't get enough thought time or air time. We all want to feel competent, to be connected to others and to have choice in our lives. Rational analysis is subservient to those goals. Inconveniently from the point of view of the challenges laid out in the first half of the book, we aren't rational beings who automatically act on evidence and honour truth.

Ice

Stuff that is melting to the point of extinction, in its natural form, from increasing chunks of the Earth's surface.

Individualism tent

Fashionable on both sides of the Atlantic and elsewhere. A fundamental tent of neoliberalism and free market economics. However, there is a logical disconnect between individualism and the treatment of global challenges. The former is inadequate for the latter.

Inequality

Too high and getting worse. Correlates inversely with most well-being measures. Big inequality is a key inhibitor of global governance, particularly when it comes to anything that involves agreeing the sharing out of the world's resources.

Investment

Ever since humans started holding back from eating all their food, so that some could be planted for next year, investment has been the means by which we have grown the economy. What we invest in determines the future we grow into. Everything we spend money on supports one kind of future at the expense of others. This applies to a simple trip to the shops, to your pension portfolio, to the national budget.

Since there is only so much money to go around, divestment liberates the opportunity for investment elsewhere. Just to give one important example, investment in fossil fuels is unhelpful. However, divestment from them creates opportunity for investment in things we urgently need like renewables, green transport and direct capture of carbon from air.

IPCC

Intergovernmental Panel on Climate Change. A large group of very clever and well-resourced scientists who have put together an incredibly thoughtful and robust assessment of climate change; the risks and uncertainties. In the past it has had two main shortcomings. The first has been the misplaced belief that presenting the arguments ever more clearly with increasing rigour is what is required to win the argument and precipitate appropriate action on climate change. (See Human denial, Bullshit, Fake news, and Chapter 8 – Values, Truth and Trust.) The second shortcoming has been that in an attempt to avoid criticism, it has been inclined to underplay some of the more uncertain climate change risks. Its 2018 report, 'Global Warming of 1.5 °C', makes its most compelling call so far for urgent global action.

Jobs

A way of spending time that can be useful, fulfilling and which can be a mechanism for appropriate wealth distribution. Worth having when at least two of these three criteria are met, but otherwise not. Therefore to be used with caution as a national performance metric.

Kids (ours)

The people who will have to understand better than their parents the nature of the Anthropocene challenge and how to deal with it.

Leadership

A scarce and much needed quality for dealing with the issues covered in this book. Anyone can display it in any walk of life and small actions can occasionally go viral. Pitifully lacking among most politicians worldwide as I write this, but can be encouraged by simple carrot-and-stick training by voters.

Local food

Sometimes a very good idea but not always. Shipping is efficient enough to make food from all over the world acceptably sustainable. It is OK for people in cold climates to eat oranges, bananas and rice. It is much less OK for them to eat hot-housed local vegetables or local meat fed on soybeans that have been grown on the other side of the world.

Maldives

Among the first and most obvious victims of climate change. The Maldives makes the problem less abstract but, for most of us so far, still not sufficiently visceral to inspire the required response.

Maxwell–Boltzmann distribution

The way energy gets distributed between atoms in a gas. Human wealth distribution looks a bit like this but in most countries the similarity breaks down for the very rich. Whereas very fast atoms usually pass on some of their energy when they collide with others, very rich people usually get even richer when they have a financial interaction with poorer people. So wealth starts to beget wealth. How can people be more like atoms?

Meat and dairy

When animals are fed human-edible crops, this is the most inefficient part of the global food supply in terms of calories, protein and micro-nutrients. The inefficiency is also costly in greenhouse gas terms especially when the animals are ruminants (sheep and cattle) and typically double their impact through the belching of methane. Other problems include the two thirds of the world's antibiotics that are used in animal farming, and the welfare of the beasts themselves, and, with much of the world's farming, the risks of nasty disease mutations.

Media

Critical in helping us understand what is going on, provided it maintains a clear distinction between fact and fiction. To this end we need to be more demanding and critical of it. Choose what you watch, read and support the best media with your money with care. (See Chapter 8 – Values, Truth and Trust.)

Methane

In the phrase 'coal, oil and gas', methane is the gas we are talking about. As well as being commercially extracted, it is also produced by cows and sheep when they chew the cud (ruminate), by paddy fields if they are flooded, it pours out of permafrost if it melts, and is produced by rotting organic matter in landfill sites. When burned it releases a lot of heat energy (that's good) but produces carbon dioxide (that's bad). Much more problematic is that if it is released into the atmosphere without burning it is a much more powerful greenhouse gas than carbon dioxide. In fact if you compare the global warming impact of methane over 20 years to the same weight of carbon dioxide of that period you find it is 76 times worse. However, methane only has a half-life of about 12 years in the atmosphere, so over a 100-year period, the slow acting but long-lasting carbon dioxide has caught up a bit and the methane is 'only' 25 times worse. There is an arbitrary convention to take a 100 years' timeframe, for most purposes. However, there is a very strong case for being interested in much shorter timescales than this, in which case methane needs to be considered as an even more powerful greenhouse gas.

Neoliberalism

The idea that the free market is the best way of running the world coupled with the flawed idea that making the rich richer helps the poor as well (trickledown). This book has repeatedly shown the inadequacy of the free market in dealing with Anthropocene challenges and trickledown has also been debunked. Neoliberalism belongs in the bin.

Neuroscience

Rapidly emerging science of brain chemistry that has reassuring messages about the human mind's plasticity; our capacity to learn to think in different ways, including, for example, the ability to adopt Anthropocene-appropriate values and thinking skills outlined in this book.

Nuclear power

A source of energy that is continuous, reliable, expensive, risky, permanently polluting, and unsurprisingly contentious. The extent of its rightful place in the world energy mix depends on a highly complex analysis of these factors compared to those associated with having to find the energy from elsewhere. A fossil plus nuclear future is the worst of all worlds. A renewable plus nuclear future could be better than any future with fossil fuels. (See Is nuclear nasty? on page 85.)

Nuclear fusion

The theoretical promise of unlimited energy has tantalised scientists for many decades. Whether this would solve all our problems or be a total disaster depends on whether you'd trust our species with yet more energy, or whether you think we do enough damage with the existing supply.

Ocean acidification

Nasty and inadequately understood consequence of burning too much fossil fuel. At its most obvious, this threatens fish stock collapse. Other consequences could be appallingly far-reaching for land-based life too. Very hard to reverse once it has happened.

Oil

Along with coal and gas, one of three substances that we need to leave in the ground. Its energy density and fluidity make it the easy choice for powering cars, ships and, perhaps most critically, planes. But its time has to end fast.

Optimism bias

A human tendency towards thinking that things will be better than they are. Most of us think that our kids are smarter than average, that we will earn more than average and live longer than average. And for some this applies to human survival in the Anthropocene: 'We will be fine because we always have been.'

There are two problems with this line of argument. Firstly, humans have inflicted some very hard times on themselves in the past, sometimes wiping out whole communities. Now that we are a global village, like it or not, it would be a great shame if our community got into as much trouble as the people on Easter Island who chopped down the last tree. The second problem with this line of reasoning is that it actually says nothing at all, because every member of every species that ever lived could say, during their lifetime, that their species had always survived. Is it optimism bias that tells me we have everything to play for?

Parents
People (including me) who now have a responsibility to help their kids develop the thinking skills to crack the Anthropocene challenges that they themselves have been unable to get to grips with.

Personal actions
Depending on how and what we all do, these could be the route to all change or totally futile. How can we know the difference? How can we know whether our personal actions are getting lost in the balloon squeezing rebounds of the global system, or are helping to create the conditions for humanity to become ready to transform into an Anthropocene-fit mode of existence? It is the wider ripples of our actions that matter most.

Plastic
Wonderfully useful, conveniently cheap and devastatingly durable. Yet another example of a great invention that has been used without enough care. Most gets discarded to landfill or scattered across land and sea, where it falls apart but does not dematerialise. So, we are stuck with it for all time once it is out there. (See pages 62–65.)

Population
Not the root of all our problems, as some people think: 12 billion careful people could live well on the planet Earth, whereas 1 billion

careless people couldn't. (See Does it all come down to population? on pages 169–171.) A smaller population would probably make it easier, though.

Prison

Usually a very nasty place to spend time, the likelihood of which depends vastly on your nationality, ethnicity, gender and wealth. Sometimes run for profit, sometimes for revenge, sometimes to keep the public safe. In a few enlightened countries, these are humane places where rehabilitation is of primary concern. The result is higher costs per prisoner but lower prison and crime costs overall. Does a nation express its values through its prisons? Perhaps. How depressing for the USA and UK. (See What are my chances of being in prison? on pages 175–178.)

Refugees

People we will need to treat entirely differently once Anthropocene-fit values become prevalent. People we will hopefully see fewer of, once global inequalities have been reduced to a level that allows the global cooperation that Anthropocene challenges demand.

Rebound effects

See Balloon squeezing. A key concept for understanding global system dynamics and one that climate policy makers all need to get their heads around, but too often find too inconvenient to contemplate. (See for example pages 91–93, 108–112 and 229 onwards.)

Renewables

Tomorrow's energy supply. Dominated by solar energy but significantly backed up by wind, tidal and hydro. Whether it augments or replaces fossil fuel will depend on whether there is a firm cap on the use of fossil fuel.

Rice

Cereal crop responsible for 19% of global calories, and 13% of the world's protein. There is an opportunity to save an estimated

500–800 million tonnes of CO_2e, simply by being more judicious with the use of fertiliser and avoiding the flooding of paddy fields.[5]

Russia

The largest country in the world. The first impacts of climate change will include unfreezing its ports, increased fertility of some of its land, newly accessible oil and gas reserves and milder winters. The world still needs Russia wholeheartedly behind a global climate deal. One illustration of the need for radically improved international relations.

Shipping

The means by which the majority of the world's food and freight tonne miles take place. The source of 3% of the world's greenhouse gas emissions in exchange for 90% of the world's international trade of products. Long-distance shipping is tricky to electrify because it requires a lightweight form of energy storage. So as for aviation (see Aeroplanes), we may have to stick with liquid hydrocarbons for the foreseeable future. They can't come out of the ground, and biofuels put too much pressure on food security and biodiversity, so the best bet is probably to manufacture them using solar power.

Shock

One mechanism for waking up. Some people say it is the only one humans are capable of using. This had better not be true because there is a long lead time on turning around environmental problems. So if we wait for the first seriously shocking symptoms we will end up having a very uncomfortable time indeed. The good news is that just because we have usually needed big shocks in the past doesn't mean it is a forgone conclusion that we need to have a big shock now. (See also Evolution and Determinism.)

Perhaps more dangerous is our *lack* of shock, our gradual acceptance of a degraded world and declining standards of truth. George Monbiot calls it 'shifting baseline syndrome' and Charles

Handy used the analogy of the frog in a slowly heating pan of water that doesn't notice it is being cooked.[6]

Solar power

The great renewables breakthrough that stands to make the transition from fossil energy possible, albeit with wind, hydro and perhaps nuclear in supporting roles. Solar stands to make deserts productive and provides a revenue alternative for several oil-dependent countries. The danger is that we simply have the solar as well as the fossil fuel, rather than using it as a replacement.

Soya beans

An extremely nutritious crop that can also be made very tasty. Kilo for kilo, the simple soya bean beats beef in almost every one of the nutrients that are essential for human health: calories, protein, including every amino acid, vitamins, minerals, fibre. However, soya is often grown as cattle feed, despite the fact that cows convert less than 10% of the calorific value of the soya bean into beef, and are similarly inefficient with the other nutrients that soya gives them. The practice of feeding soya beans to cows, therefore, is basically bonkers. Soya as a feed crop puts a pressure on land that has been responsible for considerable deforestation. But don't blame the innocent soya, blame the cattle farms.

Space travel

The route to Planet B? For wishful fantasists, this is the means by which humans can carry on expanding as usual. For the very rich and environmentally reckless, there is tourist potential to spend time in a capsule in orbit or perhaps one day travel to Mars. For thoughtful realists, space travel is sci-fi for the foreseeable generations. (See page 133.)

Spirituality

If there is a limit to the scope of this book, this surely has to be it. But is that good enough? From a purely practical point of view, can we get by without it? Perhaps it is the essential means by which we develop the thinking skills that I have said humans urgently need. Sometimes scoffed at by scientists, sometimes unfashionable and not always strictly logical. I'm not qualified to say more on this but call for respect for any who might be able to.

Sticking plasters (band aids)

A term to describe actions that deal with symptoms but not the underlying issues. Essential sometimes but hopelessly insufficient for us in the Anthropocene. An A to Z of issues that might need sticking plasters might start off with antibiotic resistance, biodiversity, climate change, disease outbreak, eutrophication, food shortage, germs, ... It is a very big list. Meanwhile an A to Z of issues that sticking plasters can't reach might include awareness, culture of truth, global empathy, equality, critical and systems thinking, rethinking growth, respect, truth ...

Takeaways (takeouts), fast food and ready meals

All of these food forms have picked up bad reputations but could be tasty, healthy, convenient, environmentally friendly ways of providing food that can be tailored to almost any budget. We just have to do them well. Street food done well is great, particularly for those who don't own a saucepan, don't know how to use one or for whatever reason are trying to keep life's daily chores to a minimum. The packaging problem needs to be cracked.

Tax

A key mechanism for enabling a better future. A way of discouraging 'bad things' by making them more expensive, and also creating a pot of money to spend on 'good things'. 'Bad things' might include the burning of fossil fuels. Good things might include subsidies for alternatives to fossil fuel, health and education systems and other infrastructure for enabling all people in society to have healthy and fulfilling lives, and compensation for countries that rely on fossil fuels. Taxes don't in themselves make the people of a country either richer or poorer, because the amount of money taken is equal to the amount of money given back out in one form or another. Nor do they, on balance, inhibit freedom.

Tax often gets a bad name, probably because the down side of it, the bit where the money gets taken away, is usually more visible than the good side of it; all the great ways in which that money might have been spent or even paid back out to us, as well as the benefits we get from all the tax that others have paid.

Technology

Something we have been developing continuously over the millennia, and essential to our arrival in the Anthropocene. Do we drive it or has it been driving us? A key challenge for our time is can we now take control of it, use it where it helps us and learn to let it go when it doesn't.

Trickledown

The idea that if the rich can get richer, everyone wins because the wealth trickles down. A popular concept with free marketeers (see the discussion on neoliberalism, page 147), and with many of the very rich. But does it work? No.

Truth

The complexities of life in the Anthropocene make 'truth' easier than ever to distort. Those same complexities make truth an even more precious thing. We need the clearest view of reality that we can possibly get. How can we create a culture that expects and insists on truth at every turn? How can we all get better at spotting the real thing from the fake? An acupuncturist working on the whole multi-disciplinary Anthropocene challenge would take 'culture of truth' as one of the top pressure points.

Two degrees

This previously agreed 'safe limit' for temperature rise above pre-industrial levels is no longer considered to be safe. This is why the 2015 Paris talks agreed to 1.5 °C if at all possible. The 2 °C limit was based on risk assessments dating back to the 1990s that have now been superseded with better science. It now looks very risky, not least because of the likelihood of triggering irreversible positive feedback mechanisms, which would then lead to temperature increases well beyond 2 °C.

Value of human life

Is this universal or does it vary between people? Can you put a financial value on it? Whether or not we like the idea, that is exactly what we do all the time. In the UK, the National Health Service can justify a treatment if it can save a life year per £20,000–30,000.

As a rule of thumb, many economies place the value of human life at about the level of the average lifetime earning capacity of someone in that country. Valuing life by its earning capacity is inconsistent with the values I'm writing from or prescribing.

Violent death

In 1994, the Rwandan genocide pushed the global violent death rate up to 2%, but even with this, on average since 1990 just under 1% of humans have died from murder, conflict, terrorism or state execution.[7] This is an astonishingly low figure compared to the twentieth century average of around 4%.[8] It is not clear, however, whether the recent downturn is an indication that we are learning not to do each other in – in which case we might hope that the trajectory can continue – or whether, more alarmingly, is it a sign that we are at risk of a pendulum swing in the other direction. Are we in an especially dangerous period in which, because few of us can remember the terror of war, we have become careless with the peace? Time may tell. And perhaps we can influence it too, especially by addressing extreme inequality.

Voting

The ability to do this is what we call democracy, but under what circumstances is it a good idea to vote? In the UK, in the past 100 years at least, not one parliamentary seat was determined by a single vote. So if all you are voting for is to further your own interests, it turns out to be an incredibly inefficient way of spending minutes of your time, perhaps equivalent to working for a few pence an hour. Better to spend the time feathering your own nest in some other way, surely, if that is your objective. Only a fool would waste time voting for their own self-interest. But if you are voting for the best interests of the millions of people affected by the election then your vote becomes millions of times more valuable, and may be the most important thing you do that day, month or year. Smart people either vote with the common good in mind or stay at home.

Waking up
One of the critical keys to the Anthropocene challenge. A frog in a saucepan of water doesn't notice the temperature going up until it's too late. We've noticed what's happening – kind of – but our slumber is still much closer to sleep than consciousness. Ever had a dream where you want to wake up but have to fight your way out through the layers? That's where we are now.

Waste food
Second to meat and dairy, waste is the biggest source of inefficiency with the global food system. There are losses at every stage in the supply chain from field to fork. The biggest areas are harvesting losses, storage losses in less developed countries and consumer waste in developed countries. In total, globally we lose over 1,000 calories per person per day (see pages 40–49).

Wellbeing
I was recently at an event at which a respected philosopher poked fun at a new fad of people pursuing wellbeing. An audience of several hundred tittered at the banality of such a superficial quest and left me wondering what I was missing. Let's just say that the wellbeing of people and planet has to be a better goal than GDP growth.

NOTES ON UNITS

This book is full of calculations and numbers, some 'back of the envelope' and some more careful and detailed. I've tried to relate things to everyday comparisons where possible, most of the numbers have units attached to them, and so might not mean much unless you know your way round those units. I hate being told that an earthquake scored 7 on the Richter scale, because I have no grasp of whether that is a big number in that context. What's the use of knowing a statistic in hectares or terrawatts or barrels if you have no idea how big any of those things are? This section is going to try to put that straight.

Power and energy

One kilocalorie (kcal) is 4.2 kilojoules (kJ) and is the energy it takes to heat a litre of water by one degree centigrade (°C). So the average human daily food requirement of 2,353 kcal is about the amount required to run a large hot bath.

One joule (J) is the amount of energy it takes to lift a 1 kg bag of sugar 10 cm up in the air. It can heat up a large rain drop by about 1 °C (degree sign) or power a small torch for a second.

One watt (W) is a *rate* of use (or generation) of energy. 1 W is 1 J per second. It takes about 20 W to light a reasonably large household room with light-emitting diodes (LEDs).

One kilowatt (kW) is a thousand watts. A typical kettle runs at 1.8 kW.

One kilowatt hour (kWh) is the energy you use if you have a power of 1 kW for an hour. That is very roughly enough to drive an efficient electric car for about four miles or a conventional petrol car for one mile. Or you could have the kettle on for 33 minutes. 1 kWh is 3.6 MJ (that is 3.6 million joules).

One megawatt (MWh) could get you from Los Angeles to New York in an efficient car, although a 'gas-guzzler' would conk out somewhere near the California state border.

A gigawatt (GW) is a billion watts.

A terrawatt (TW) is a trillion watts (that's 12 zeros). In an average moment, humankind is using energy at a rate of almost 20 TW.

One terrawatt hour (TWh) is roughly the kinetic energy of my bike (without me) travelling at a tenth of the speed of light (which would be London to New York in half a second). Does that make a TWh feel real enough? Perhaps not. Try this. It is enough energy to bring 580 Olympic swimming pools to the boil and then boil them dry.[1]

(Zettawatt, yottawatt and exawatt all have even more zeros and so I don't use them in this book.)

Distance

Miles: Old fashioned, but most of the English-speaking world still uses them instead of kilometres. 1 mile = 1.61 km, so 1 square mile = 2.6 km^2.

Greenhouse gas emissions

kg CO_2e – kilograms of carbon dioxide equivalent. This is a crude way of combining all the different greenhouse gases, especially carbon dioxide, methane and nitrous oxide, into one measure.[2] Sometimes in this book I use 'carbon' as a shorthand for 'greenhouse gases'.

Humanity causes roughly 50 billion tonnes CO_2e per year. In kilograms that is 5 and then 13 zeros. Each of the following has a carbon footprint of roughly 1 kg CO_2e: just over 1 kWh of Chinese or Russian electricity, or burning 0.4 litres of petrol. The average UK person has a carbon footprint of around 15 tonnes CO_2e per year – that's if you count all the greenhouse gases that lie behind everything they do and buy in a year, and the average US citizen almost double that.[3]

Weights

One kilogram is 2.2 pounds (lb). Whenever I talk about tonnes, I mean the metric ones: 1,000 kg or 2,200 lb. A litre of water weighs a kilogram, so a cubic metre of water weighs a tonne.

Stuff I don't use

I've steered clear of hectares (which are 100 m × 100 m, so there are 100 of them in a square kilometre) and acres which are just over half the size of hectares. I wish no one would ever mention barrels of oil or worse still barrels of oil equivalent, because almost nobody understands how big these are.

ENDNOTES

What's New in this Updated Edition?

1 IPCC (2018) Global Warming of 1.5 degrees C – Summary for Policy Makers. http://report.ipcc.ch/sr15/pdf/sr15_spm_final.pdf

Introduction

1 For the tsunami energy estimate of 1×10^{17} J, thanks to Wikipedia for a delightful list of energies by order of magnitude: www.wikipedia.org/wiki/Orders_of_magnitude_(energy). This particular one comes from 'USGS Energy and Broadband Solution', National Earthquake Information Center, US Geological Survey. (Archived from the original on 4th April 2010. Retrieved 9th December 2011.)

For global energy use, data come from the BP statistical review 2017. I have not included the energy in human food eaten here, although later in the book I do include this as part of the human energy supply. It is around 5% of the total.

2 From the Q&A session after Stephen Hawking's second Reith lecture in 2016. You can read the full transcript here: https://tinyurl.com/ReithHawking

3 *'There is no Planet B'*. Thanks to Emanuel Macron for getting this useful phrase into the headlines in 2018. It was originally coined, I think, in 2011 by José María Figueres, former Costa Rican president and later used in 2014 by UN Secretary General Bhan Ki-Moon. It is also the name of an international primary school project to tackle climate change. www.theresnoplanetb.co.uk/

4 I'm not by any means the first person to spot this. For example, see Ken Wilbur, *The Theory of Everything: An Integral Vision of Business, Politics, Science and Spirituality*, Shambhala, 2000.

1 Food

1 2,353 kcal per day (9.9 MJ) is the Average Daily Energy Requirement (ADER), as estimated by the UN Food and Agriculture Organization as a weighted average across all ages, genders, weights and lifestyles.

2 An electric kettle is typically 1.8 kW.

3 M. Berners-Lee, C. Kennelly, R. Watson and C. N. Hewitt (2018) Current global food production is sufficient to meet human nutritional needs in 2050 provided there is radical societal adaptation. *Elementa: Science of the Anthropocene.* https://tinyurl.com/LancFoodPaper. This paper contains the global food analysis referred to extensively in this section, as well as a great deal more detail. The supplementary information has an Excel spreadsheet to allow the assessment of flows of most human-essential nutrients at the global, regional and national level. (Although beware that data quality often breaks down at the national level.) It took a lot of work!

4 See note above.

5 A. Saltzman, K. Von Grebner, E. Birol et al. (2014) Global Hunger Index: The Challenge of Hidden Hunger. International Food Policy Research Institute, Bonn/Washington DC/Dublin. Also S. Muthayya, J. H. Rah, J. D. Sugimoto et al. (2013) The global hidden hunger indices and maps: an advocacy tool for action. *PLoS One* 8, e67860.

Iodine shortage can be solved through supplements at a cost of something like $0.02 to $0.05 per person per day.

6 M. Berners-Lee, C. Kennelly, R. Watson and C. N. Hewitt (2018) Current global food production is sufficient to meet human nutritional needs in 2050 provided there is radical societal adaptation. *Elementa: Science of the Anthropocene.* https://tinyurl.com/LancFoodPaper

7 One kilogram (1 kg) of human body weight requires about 7,700 kcal of food in excess of energy used. So 100 kcal per day in excess of requirement would lead to 36,500 excess calories per year. Almost 5 kg per year (4.74 to be exact), or 95 kg in 20 years.

8 330 kcal per person per day is the net excess, after taking account of the people who go hungry. The excess consumed by those who eat too much probably equates to 500 kcal per person per day for everyone alive today. That is enough to feed more than a fifth of the current population of roughly 7.5 billion. So my 1 billion figure is erring on the underside.

9 Note that there is potential for negative feedback here as well. Better diets would increase life expectancy and contribute to the population growth.

10 M. Berners-Lee, C. Kennelly, R. Watson and C. N. Hewitt (2018); see note 6 above.

11 Sweet potatoes contain 709 micrograms (μg) of vitamin A per 100 g according to United States Department of Agriculture (Food Composition Databases (2017). http://ndb.nal.usda.gov, accessed 13th September 2017), compared to a recommended daily intake of 900 and 700 μg for men and women. Data on sweet potato production in China come from the Food and Agriculture Organization of the United Nation Statistics Division Food Balance Sheets (2016). http://faostat3.fao.org/browse/FB/FBS/E, accessed 13th September 2017.

12 Iodine is also important but supplements are so cheap (~$0.02 to $0.05 per person per day) that I have left them out of the discussion.

13 Vitamin A content of a few foods (source USDA, see note 11 above):

	Micrograms vitamin A /100 grams
Sweet potato	709
Olives	26
Maize	11
Tomatoes	38
Green leafy vegetables	95
Pimento	1,524
Poultry meat	173
Butter	684
Eggs	160

14 There is a lot of uncertainty surrounding global antibiotics use due to the lack of effective monitoring, but estimates from F. Aarestrup (2012) Sustainable farming: get pigs off antibiotics. *Nature* 486 (7404), p. 465 and a World Health Organization report titled 'The evolving threat of antimicrobial resistance, Options for Action' (available from: https://tinyurl.com/reportwho) suggest that antibiotic use in animals is as much as twice that in humans. The *Guardian* puts it even higher, at nearly 75% of antibiotics worldwide used on animals. https://tinyurl.com/farmingantibiotics

15 T. P. Van Boeckel, C. Brower, M. Gilbert et al. (2015) Global trends in antimicrobial use in food animals. *Proceedings of the National Academy of Sciences* 112(18), pp. 5649–5654.

16 World Health Organization Media Centre, Antibiotics Resistance Fact Sheet. https://tinyurl.com/whoantibiotics

17 Soil Association standards on animal welfare. https://tinyurl.com/organicanimals

18 Rob Wallace (2015) Big farms make big flu: dispatches on influenza, agribusiness, and the nature of science. *Monthly Review*, available from: https://monthlyreview.org/product/big_farms_make_big_flu/

19 Is factory farming to blame for coronavirus? *The Observer*, March 2020, available from: https://bit.ly/2Uqc4hl

20 The figure of 23% was our estimate in *The Burning Question*, and the derivation is in the endnotes. More recently, a major paper (and highly recommended) in the journal *Science* puts the food supply chain at 26% (13.7 billion tonnes per year) and another 5% (2.8 billion tonnes per year) associated with non-food crops. See also Mike Berners-Lee and Duncan Clark, *The Burning Question*, Profile Books, 2013 and the 2018 article by J. Poor and T. Nemecek in *Science* (see next note).

21 J. Poore and T. Nemecek (2018) Reducing food's environmental impacts through producers and consumers. *Science* 360(6392), pp. 987–992. In almost all cases, the results are strikingly similar to my own research, including my work for supermarkets and the research that went into one of my previous books, *How Bad Are*

Bananas? It is not surprising because most of the sources are the same. The biggest difference is that Poore and Nemecek pull out the significance of land-use change in a way that I had not properly cottoned onto when I wrote *Bananas*. *Bananas* goes into a lot more detail about some foods, especially fruit and veg, and the effect of seasonality, local and imported.

22 See George Monbiot's piece 'Fowl Deeds'. https://tinyurl.com/MonbiotChickens. See also BBC Countryfile: Free-range poultry farming contributing to increase in river pollution, warn authorities, October 2017. https://tinyurl.com/yayk3l66

23 Mike Berners-Lee, *How Bad Are Bananas? The Carbon Footpint of Everything*, Profile Books, 2011. Revised edition, 2020. M. Berners-Lee, C. Hoolohan, H. Cammack and C. N. Hewitt (2012) The relative greenhouse gas impacts of realistic dietary choices. *Energy Policy* 43, pp. 184–190. *Energy Policy* 2011. http://dx.doi.org/10.1016/j.enpol.2011.12.054 and C. Hoolohan, M. Berners-Lee, J. Mckinstry-West and C. N. Hewitt (2013) Mitigating the greenhouse gas emissions embodied in food through realistic consumer choices. *Energy Policy* 63, pp. 1065–1074. http://dx.doi.org/10.1016/j.enpol.2013.09.046

24 Creating a Sustainable Food Future, Installment Eight, from the World Resources Institute, covers water management in more detail. https://tinyurl.com/globalriceGHG

25 The greenhouse gas footprint of Booths (2015). https://tinyurl.com/ghgbooths

26 M. Marschke and P. Vandergeest (2016) Slavery scandals: unpacking labour challenges and policy responses within the off-shore fisheries sector. *Marine Policy* **68**, pp. 39–46. https://tinyurl.com/Fishingslavery

27 Also B13, D3, potassium and omega 3 fatty acids. But the minerals listed in the main text are a more critically important contribution of fish to nutrition in the developing world.

28 A. D. Rijnsdorp, M. A. Peck, G. H. Engelhard et al. (2009) Resolving the effect of climate change on fish populations. *ICES Journal of Marine Science* 66(7), pp. 1570–1583. https://tinyurl.com/climatefish

29 Marine Conservation Society. https://www.mcsuk.org

30 The sad possibility that the Marine Stewardship Council might be fatally flawed by financial vested interests is articulated in the following article in the Fish Information & Services (FIS) website: On The Hook: UK supermarkets caught in 'unsustainable tuna scandal'. https://tinyurl.com/marinesc

31 J. L. Jacquet and D. Pauly (2007) Trade Secrets: Renaming and mislabelling of seafood. *Marine Policy* 32, pp. 309–318. Available from: https://tinyurl.com/seafoodlabels

32 For more detailed guidance on how to approach your fishmonger go to Wetherell, 2018: The sustainable food trust. https://sustainablefoodtrust.org/articles/eating-values-five-questions-ask-fishmonger/

33 www.mcsuk.org/goodfishguide. Huge apologies to Marine Conservation Society (MCS), a wonderful NGO that I confused with the MSC in the first edition of this book. And thanks to several readers for pointing out my error.

34 For detailed breakdowns of calories by supply chain stages and regions, see M. Berners-Lee, C. Kennelly, R. Watson and C. N. Hewitt (2018); note 6 above.

35 J. A. Moult, S. R. Allan, C. N. Hewitt and M. Berners-Lee (2018) Greenhouse gas emissions of food waste disposal options for UK retailers. *Food Policy* 77, pp. 50–58.

36 Based on 70% methane capture, which is about as good as landfill ever gets. See paper referenced above.

37 718 kcal for a Sainsbury's stonebaked margarita pizza. http://tinyurl.com/gta4ryx

38 If in doubt about whether lab meat is a more attractive concept than today's meat industry, this 6-minute video from SAMSARA may help you make up your mind: https://vimeo.com/73234721

39 The Bill & Melinda Gates Foundation awarded the International Rice Research Institute (IRRI) funding for its C4 Rice project in 2008. The project is currently on Phase III (2015–2019). More information is available here: https://c4rice.com/

40 See for example: Protein produced from electricity to alleviate world hunger, LUT University, available from: https://bit.ly/

2UluQWf. Thanks also to Channel 4 and George Monbiot for an interesting piece on this in their 2020 documentary, Apocalypse Cow: How meat killed the planet, available from: https://bit.ly/3aVEIMU

41 Data drawn from: M. Berners-Lee, C. Kennelly, R. Watson and C. N. Hewitt (2018); endnote 6.

42 The UN projected that the global population will rise to 9.7 billion by 2050 in its 2015 World Population Prospects report. https://tinyurl.com/UNworldpop

43 First-generation biofuel comes from human-edible crops, second generation from other feedstocks, typically cellulose-based.

44 For more detail, The Convention on Biological Diversty has 20 targets. www.cbd.int/sp/targets/

2 More on Climate and Environment

1 The Convention on Biological Diversity report 'Sustaining life on Earth' (2000) settles on 13 million (https://tinyurl.com/cbdreport). More recent studies estimate around 8 million (C. Mora, D. P. Tittensor, S. Adl, A.G. Simpson and B. Worm (2011) How many species are there on Earth and in the ocean? *PLoS Biology* 9(8), e1001127. https://tinyurl.com/totalspecies), although it is acknowledged that estimates up to 100 million species could be defended (R. M. May (2010) Tropical arthropod species, more or less? *Science* 329(5987), pp. 41–42. https://tinyurl.com/speciesscience).

2 According to the World Wildlife Fund estimates, available here: https://tinyurl.com/wwfbio

3 Estimates come from the International Union for Conservation of Nature's Red List, which stresses that this may be partially down to changes in species classification as well as in the definition of 'threatened'. The red list is available here: https://tinyurl.com/redlistiucn

4 World Wildlife Fund (WWF), Living Planet Report 2016. https://tinyurl.com/wwflivingplanet

5 C. A. Hallmann, M. Sorg, E. Jongejans et al. (2017) There has been more than a 75 percent decline over 27 years in total flying insect

biomass in protected areas. *PLoS One* 12(10), e0185809. https://doi.org/10.1371/journal.pone.0185809

6 B. J. Cardinale, J. A. Duffy, A. Gonzalez et al. (2012) Biodiversity loss and its impact on humanity. *Nature* 486(7401), p. 59. This *Nature* paper draws on hundreds of other pieces of research. 'For provisioning services, data show that (1) intraspecific genetic diversity increases the yield of commercial crops; (2) tree species diversity enhances production of wood in plantations; (3) plant species diversity in grasslands enhances the production of fodder; and (4) increasing diversity of fish is associated with greater stability of fisheries' yields. For regulating processes and services, (1) increasing plant biodiversity increases resistance to invasion by exotic plants; (2) plant pathogens, such as fungal and viral infections, are less prevalent in more diverse plant communities; (3) plant species diversity increases aboveground carbon sequestration through enhanced biomass production (but see statement 2 concerning long-term carbon storage); and (4) nutrient mineralization and soil organic matter increase with plant richness.'

7 An interview with Jane Lubchenco in the Yale Environment 360 online magazine can be found here: https://tinyurl.com/janelubchenco

8 The pH of the ocean has decreased by 0.1 since the Industrial Revolution. This may not sound like much, but it corresponds to at least a 26% increase in the hydrogen ion concentration (or acidity) in seawater, and this is projected to rise to 150% by 2100 (see Chapter 3 in IPCC (2014) *Climate Change 2013 – The Physical Science Basis*, Working Group I Contribution to the Fifth Assessment Report of the Intergovernmental Panel on Climate Change, Cambridge University Press). Furthermore, scientists believe the rate of acceleration of acidification is greater than any other ocean geochemical changes in the past 300 million years, meaning it is unprecedented and we could be heading into unknown territory when it comes to marine ecosystem change (B. Hönisch, A. Ridgwell, D. Schmidt et al. (2012) The geological record of ocean acidification. *Science* 335(6072), pp.1058–1063).

9 Quoted in an archived report from *National Geographic*, available here: https://tinyurl.com/ThomasLovejoy
10 All of the numbers on plastic production and fate, unless otherwise stated, come from R. Geyer, J. R. Jambeck and K. L. Law (2017) Production, use, and fate of all plastics ever made. *Science Advances* 3(7), e1700782. https://tinyurl.com/fate-of-plastic
11 Enough to wrap Earth 1.33 times according to my sums. The surface area of Earth is 510 million km^2. Assuming a thickness of approximately 8 μm, which is what you will find for most household cling film, the volume of cling film needed to wrap the world would be 4 billion m^3. The volume of plastic discarded up until 2015 was 5.4 billion m^3 (assuming in this case it all has a density of cling film: 0.92 g/cm^3) meaning Earth can be wrapped 1.33 times, leaving enough spare for a double thickness over the land, or to also wrap the moon four times over.
12 J. R. Jambeck, R. Geyer, C. Wilcox et al. (2015) Plastic waste inputs from land into the ocean. *Science* 347(6223), pp. 768–771. They made their estimates of plastic waste entering the ocean by linking worldwide data on solid waste generation, population density and quality of waste management.
13 According to A. L. Lusher, M. McHugh and R. C. Thompson (2013) Occurrence of microplastics in the gastrointestinal tract of pelagic and demersal fish from the English Channel. *Marine Pollution Bulletin* 67(1), pp. 94–99. https://tinyurl.com/plasticinfish
14 Note that while breaking down into small pieces, which is sometimes described as biodegrading, might take only a few hundred years, this is not the same as mineralisation; the actual breaking down of the molecules themselves, which usually does not happen at all in the ocean.
15 UK Government, Foresight Review of Evidence: Plastic Pollution (2017). https://tinyurl.com/foresightplastic, in turn referencing J. R. Jambeck, R. Geyer, C. Wilcox et al. (2015); see note 12 above.
16 The amount of plastic being thrown into landfill or the natural environment per year will continue increasing for the next decade, and it will be another two decades before it returns to today's levels.

17 In 2015, the rate of oil production was around 92 million barrels per day, which is 4,561 million tonnes per year. I've assumed every tonne of oil could make 1.37 tonnes of PET (based on carbon weight), one of the most common plastics used for plastic bottles and polyester production. So if all this oil was used to make plastic we would have produced 6,255 million tonnes, equivalent to 83% of all the plastic ever made by 2015.

18 Thanks to the *Guardian* for all this section 'Is there life after plastic? The new inventions promising a cleaner world', February 2018. https://tinyurl.com/lifeafterplastic

3 Energy

1 Except for food, data come from the BP statistical review 2017. Food data and other sources which put global average consumption at 2,545 kcal per day, somewhat above the UN Food and Agriculture Organization's figure for average dietary energy requirement of 2,353 kcal.

2 Based on 65 kcal per hour which is about right for a 65 kg person according to various sources, including Captain Calculator's Calories Burned During Sleep calculator. https://tinyurl.com/sleepcalories

3 All the data apart from food come from Energy Consumption in the UK, 2018, BEIS, https://tinyurl.com/UKenergyuse; Food estimates are taken from the analysis earlier in the book.

4 See note 3.

5 BP Statistical Review of World Energy 2017 – Underpinning data. Primary energy in the form of electricity (nuclear and renewables) is converted to 'oil equivalent' based on a 38% efficiency of oil power stations. A mark-up factor of 2.6 results. https://tinyurl.com/bpreview17

6 Earth has a cross section of 1.274×10^{14} m^2 and the solar constant (sun incident per m^2, perpendicular) is 1,361 W/m^2, so 1.74×10^{17} W in total. But 30% is reflected, leaving about 953 W/m^2 or 1.22×10^{17} W total. With 7.5 billion people, this leaves 16,277 kW per person or 391,000 kWh per person per day, compared to our use of 59 kWh per person per day. So we use one 6,621th.

7 Based on a 3,000 m^3 pool and 90-degree temperature rise.

8 I have taken a typical kettle to be 1.8 kW. For the flight, I've based it on long-haul for an A380 with 500 people on board and used a clever aviation emissions model developed for me by David Parkinson at Sensus that does a pretty good job of mirroring real world fuel consumption for different plane types, payloads and flight distances. The A380 gets through about 500 kW per passenger averaged over the flight.

9 Based on cheap panels with 16% efficiency and a random spread of locations, rather than choosing the ones most suitable for photovoltaics. So this is conservative.

10 Data yet again from BP Statistical Review of World Energy 2017.

11 I write 'perhaps'. Chris Goodlall made the following comment '40% efficiency seems aggressive. Requires tandem layers, probably with perovskite.'

12 Thanks to Chris Goodall for this. For an optimistic take on the solar revolution, I recommend his book *The Switch*, published by Profile.

13 For example, see Vaclav Smil's 2015 book *Energy Transitions: History, Requirements, Prospects*, published by ABC-CLIO.

14 The calculations are made using NASA data on light intensity (https://power.larc.nasa.gov/new/; accessed 2019) and World Bank data on land area and population (https://data.worldbank.org/indicator/SP.POP.TOTL) and ARC GIS software.

15 Resource extraction for batteries: most lithium ion batteries require cobalt, much of which comes from unregulated mines in the Democratic Republic of the Congo, complete with child labour, high injury levels and mystery illness. As of 2016 Amnesty International found that no country requires its companies to report its cobalt supply chains. In terms of corporate action, in 2017 Amnesty rated Apple and Samsung 'Adequate' and everyone else below, with Microsoft, Lenovo, Renault, Vodaphone, Huawei, L&F, Tinjin, B&M, BYD, Coslight, SHenzhan and ZTE all scoring 'No Action Taken'. See Amnesty International, November 2017, Time to Recharge: Corporate Action and Inaction to Tackle Abuses in the Cobalt Supply Chain.

16 Marcus Budt, D. Wolf, R. Span and J. Yan (2016) A review on compressed air energy storage: basic principles, past milestones and recent developments. *Applied Energy* **170**, pp. 250–268.
17 Vaclav Smill, *Energy Transitions: Global and National Perspectives*, 2nd edition, Praeger, 2016.
18 Adams and Keith (2013) Are global wind power resource estimates overstated? *Environmental Research Letters*, **8** (1 March), p. 015021.
19 Based on the land area of the UK being 242,495 km^2 and the population 66.6 million.
20 This map estimates the average total kinetic energy in a band of air 50 m above ground level for each country. It is not in any way comparable with the sunlight estimates, because one is a measure of energy per day and the other is a measure of energy at any one time, before any of it is harnessed. This map ignores very important factors like terrain and offshore possibilities. Wind speed sources are from NASA: https://power.larc.nasa.gov/new/ (accessed 2019) from which kinetic energy (KE) is established from the density of air, land area and KE= $1/2\ mv^2$. Note also that I have used the square of the mean wind speed rather than the mean square – which would have been better, but I didn't have the data.
21 I estimated 19.5 TW, based on rainfall and altitude data for 50 km^2 locations and ARC GIS software. Smil (2016) in *Energy Transitions* estimated 11.75 TW. His is probably more robust, but both estimates are just to get an order of magnitude sense of hydro's potential.
22 Taking my higher figure for the potential energy in landed rain we get 19.5 TW × 5% × 80% = 0.78 TW. Smil's figure of 11.75 and the same assumptions would mean we had just about exploited all the potential already. Clearly the 5% 'catchability factor' is critical.
23 According to MIT-based Commonwealth Fusion systems, they are just 15 years away from a plant that can supply the grid. *MIT Technology Review*, 9 March 2018. https://tinyurl.com/fusion15yrs
24 The average dietary energy intake required for healthy living in the USA is 2,500 kcal per day according to the FAO, who make national adjustments, based on body size and activity levels. That is a bit higher than the global average, because Americans tend to

be larger. In the USA, more than 4,000 kcal per person per day are used for biofuels.

25 W. D. Huang and Y. H. P. Zhang (2013) Energy efficiency analysis: Biomass-to-wheel efficiency related with biofuels production, fuel distribution, and Powertrain systems. *PLoS One*. http://dx.doi.org/10.1371/journal.pone.0022113

26 For more information on algaes as a source of biofuels: M. Hannon, J. Gimpel, M. Tran, B. Rasala and S. Mayfield (2010) Biofuels from algae: challenges and potential. *Biofuels* 1(5), pp. 763–784. https://tinyurl.com/ya8zuty3

27 William Stanley Jevons, *The Coal Question*, Macmillan, 1865. Jevons pointed out that more efficiency would make coal more attractive and would increase demand rather than reduce it.

28 While just about everyone who looks at this thoughtfully is prepared to acknowledge 'rebounds' from efficiency gains, the critical question is whether or not these routinely sum to more than 100%. All the analyses I have seen that claim they do not come to their finding by placing a limit on the rebound pathways that are considered. It is actually not possible to analyse the effect of efficiency gains by looking at any one part of the economy in isolation, nor over a short period of time. The only real measure is to look at the combined effect of all the world's efficiency gains in all areas and track that against global energy use. When you do this you find a total energy rebound averaging 102.4% over the past 50 years (i.e. the annual global growth in energy use).

One of the more thoughtful papers on rebound was published by the UK Energy Research Council in 2007 (The Rebound Effect Report: https://tinyurl.com/UKERCrebound). It is 127 pages long. It categorises rebounds into various types (substitution rebounds, income/outcome rebounds, embodied rebounds and secondary rebounds). It concludes that rebounds are hugely important but that generally the Jevons Paradox does not exist because rebounds do not usually sum to more than 100%. However, this conclusion is reached because even this paper fails to take full account of all rebounds. One example of a rebound that would not be included in this study would be more efficient lighting leading to someone having a more conducive working environment, leading in turn to them having a better idea about how oil might be extracted from the ocean.

Many other studies underestimate rebound much more seriously. See next endnote.

29 For example, *SMARTer 2030*, a report by the Global e-Sustainability Initiative (GESI), available from www.smarter2030.gesi.org, claims ICT can cut global carbon by 20% by 2030, provided rebounds are not taken into account. The rebound effect is acknowledged in an appendix but the effect is grossly underestimated because macro-economic rebound pathways are ignored.

30 Most notably this is what BP do in their widely used annual statistical review of energy. This is why renewable energy generation is often measured in 'tonnes of oil equivalent' – in other words, the tonnes of oil that would be required to generate the same amount of electricity. BP uses an efficiency factor of just 38% for coal and oil power stations. Gas turbines are significantly more efficient.

31 J. Strefler, T. Amann, N. Bauer et al. (2018) Potential and costs of carbon dioxide removal by enhanced weathering of rocks. *Environmental Research Letters*, doi:10.1088/1748-9326/aaa9c4. https://preview.tinyurl.com/rockweatherring. Thanks also to Chris Goodall and his excellent weekly Carbon Commentary newsletter for this. An excellent source of low carbon technology updates with free subscription: www.carboncommentary.com/

32 J.-F. Bastin, Y. Finegold, C. Garcia et al. (2019) The global tree restoration potential. *Science* **5**, pp. 76–79. There is room on the planet for an extra 0.9 billion hectares of tree cover, supporting 517 billion trees and storing 205 Gt of carbon.

33 Boom times are back for carbon offsetting industry, *Financial Times*, December 2019, available from: https://on.ft.com/2vGGbHM

34 The scheme is run by Climeworks, available from: www.climeworks.com/

35 For example, DNV GL Energy Transition Outlook 2017 (https://eto.dnvgl.com) estimates energy demand peaking around 2050 and then falling steadily. But it is based on efficiency running at

2.5% per year (against 1.4% for the past 20 years) but assumes that demand for utility will be independent of this – in other words, zero rebound effect. It also uses the most optimistic population forecast I have ever seen (peaking at 9.4 billion; see Wittgenstein Centre for Demography and Global Human Capital, available from https://tinyurl.com/2017wittgen) and also no rebound effect on the reduction in population growth either. Even with all this they forecast temperature rise at a totally unacceptable 2.5 °C or so. Similarly, Statoil's 'Energy Perspectives 2017 Long-term macro and market outlook' has a renewal scenario in which energy demand falls by 8% to 2050. https://tinyurl.com/2017statoil

4 Travel and Transport

1 A study looking at the daily activity levels of over 700,000 people over 111 countries found that the average smart phone user walked 4,961 steps per day. The recommended 10,000 steps per day works out to be close to 5 miles/8 km – so on average we are walking 2.5 miles/4 km per day. Multiplied by total world population and 365 days in a year, we get an estimate of total distance walked in a year, for the whole population.

See T. Althoff, R. Sosic, J. L. Hicks et al. (2017) Large-scale physical activity data reveal worldwide activity inequality. *Nature* 547, pp. 336–369.

2 Passenger-km data for air travel obtained from the International Civil Aviation Organization (ICAO), rail from the World Bank and road transport from the Organisation for Economic Co-operation and Development (OECD). Converted to miles.

3 EU energy and transport in figures, Statistical pocketbook (2002).

4 European Commision (2013). Special Eurobarometer 406: Attitudes of Europeans towards Urban Mobility. https://tinyurl.com/ydyxcqa2

5 UK National Travel Survey 2017 (Department for Transport) estimated that the average UK resident cycled 60 miles per year. Boats are not included in the survey, assumed to be non-significant. https://tinyurl.com/y7tqayh6

6 A few more details. I've assumed the electricity used by bikes and cars has come from solar panels in sunny southern California. In the cloudier UK with the sun lower in the sky, we'd only get about two thirds of the mileage. For the pedal cycling, I've used the global average wheat yield and assumed that the bread you might eat to power you along can be simply made, ignoring the energy required for baking, and I've gone with a ball-park figure of 50 calories per mile and twice that for walking.

7 Public Health England (2014) Estimating Local Mortality Burdens associated with Particulate Air Pollution (https://tinyurl.com/deathsdiesel) and Royal College of Physicians (2016) Every breath we take: the lifelong impact of air pollution (https://tinyurl.com/pollutiondiesel).

8 Diesel cars emit 15 times as much PM2.5 compared to petrol and five times as much NOx. For light good vehicles, the ratios are 23 and 1.65. Data from UK National Atmospheric Emissions Inventory (NAEI). Fleet weighted road traffic emissions factors. https://tinyurl.com/emissionsroad

9 Nitric oxide (NO) is produced through a variety of natural (soil microbes, forest fires) and anthropogenic (burning of fossil fuels) processes. It is a free radical with an unpaired electron, and quickly reacts with oxygen (O_2) in the air to produce nitrogen dioxide (NO_2):

$$2NO + O_2 \rightarrow 2NO_2.$$

Alternatively, NO may react with ozone (O_3) to form oxygen and NO_2.

$$NO + O_3 \rightarrow NO_2 + O_2.$$

Nitrogen dioxide is a reddish-brown gas and is a major component of smog.

10 Drawn from Small World Consulting's analysis of the benefits of companies electrifying urban vehicle fleets. For more details, best to get in touch directly (info@sw-consulting.co.uk).

11 The 40,000 premature deaths from air pollution are thought on average to result in 12 life years lost. This is the basis on which I have arrived at the minutes per mile. I don't mean that one

person loses an hour of course, but that many people lose a moment. It's a statistical analysis. What my sums don't take into account is that because traffic pollution takes place right next to human lungs, it is disproportionately likely to cause health problems compared, say, to smoke from factory chimneys. So all my estimates are significantly, or perhaps drastically, on the cautious side.

12 Summarised in the *Independent* article: https://tinyurl.com/scandalvw. Original report: G. P. Chossière, R. Malina, A. Ashok et al. (2017) Public health impacts of excess NOx emissions from Volkswagen diesel passenger vehicles in Germany. *Environmental Research Letters* 12(3), p. 034014.

13 Here are the sums. An A380 carrying 550 passengers on the 8,062-mile journey from JFK airport to Hong Kong burns through 192 tonnes of jet fuel (using, again, our aviation simulator, built for me by David Parkinson). If the plane could be made to fly on biofuel (which in itself requires modification to the technology) that entails growing 2,166 tonnes of wheat (based on today's average efficiency of 27% of wheat to biofuel conversion and 3,390 kcal per kg of wheat). Californian wheat yields are about 0.56 kg/m^2 per year, so 1.5 square miles are needed. Human global average calorific food need is 2,353 according to the UN's Food and Agriculture Organization. So that is four years' worth of calories for each passenger.

14 I've assumed solar panels in California, lying flat with no spaces between. The average daily insolation is 5.1 kWh per m^2 per day, and I've gone with 20.4% efficiency of the panels and 85% efficiency of the energy capture to batteries.

15 Just to sense check, where does that factor of 270 come from? It is the multiplied efficiency factors of capturing the sun's energy using solar panels rather than wheat and the improved efficiency of fuel from electricity compared to fuel from wheat.

16 See my first book, *How Bad Are Bananas? The Carbon Footprint of Everything*, which also compares and contrasts numerous other travel modes.

17 A cargo freight ship from Hong Kong to London, keeping the speed down to 15 knots for efficiency, gets through only about 0.03 kWh

per mile per tonne of payload. (Derived from fuel use. N. Bialystocki and D. Konovessis (2016) On the estimation of ship's fuel consumption and speed curve: a statistical approach. *Journal of Ocean Engineering and Science* 1(2), pp. 157–166.) The journey would take two weeks.

18 David Parkinson's aviation model tells us that an A380, splitting the journey into 2,000-nautical-mile chunks for efficiency, gets through around 2 kWh per tonne per mile.

19 The *Harmony of the Seas* carried 9,000 passengers and crew and burns through 0.7 tonnes of marine fuel oil per nautical mile (which is 1.15 miles). The Department for Environment, Food and Rural Affairs (DEFRA) gives an emissions factors of 3,248 kg CO_2e per tonne for the fuel. This gives us 0.22 kg CO_2e per person per mile. None of these calculations take account of embodied carbon in the building of the boat itself. Source: https://tinyurl.com/HarmonySeas

20 For pedalos, Steve Smith's *Pedalling to Hawaii* has to be one of the top wacky adventure stories I've ever read, and for rafts, Thor Heyerdahl's *The Kontiki Expedition* is also up there. For sailing, Ellen MacArthur's *Taking on the World*.

21 For example, Steven Hawking responding to an audience question after one of his 2016 BBC Reith Lectures: 'We will not establish self-sustaining colonies in space for at least 100 years, so we have to be very careful in the meantime.' https://tinyurl.com/ReithHawking

22 Kinetic energy (KE) = $\frac{1}{2} mv^2$. $m = 50 \times 70$ kg = 3,500 kg. Speed of light (c) = 300,000,000 m/s. $v = c/10$. KE = 1.5×10^{18} J = 440 TWh or 18.3 TW for a day.

5 Growth, Money and Metrics

1 For example, Richard Wilkinson and Kate Pickerty's *The Spirit Level: Why More Equal Societies Almost Always Do Better* (Allen Lane, 2009) lays out a bulletproof case for reduction in inequality and chilling out about GDP in rich countries. *The Limits to Growth* by Donella Meadows, Jorgen Randers and Dennis Meadows (first published in 1972 by Potomac Associates-Universe Books) laid out perhaps the

first proper challenge to the 'more is better' mantra, and Tim Jackson's *Prosperity Without Growth* (Routledge, 2009) and Kate Raworth's *Doughnut Economics* (Random House, 2017) build on these themes, as does the New Economics Foundation (https://neweconomics.org).

2 For a more detailed analysis of the link between GDP (gross domestic product) and carbon emissions, see Mike Berners-Lee and Duncan Clark, *The Burning Question* (Profile Books, 2013, Chapter 9, The Growth Debate).

3 Bobby Kennedy, University of Kansas, March 1968, speaking to Vietnam War protesters. https://tinyurl.com/BobbyKennedy-onGDP

4 Tom Peters might not have been the first to coin the phrase, which is also widely associated with Peter Drucker. The idea goes back at least a century earlier.

5 The Happy Planet Index, developed by Nic Marks, measures how well individual countries are doing at achieving happy, sustainable lives. The metric incorporates four core principles: wellbeing, winequality, happiness and sustainability. http://happyplanetindex.org/

Bioregional's One Planet Living on the other hand encompasses 10 principles, with a greater breadth of environmental issues, including: health and happiness; equity and local economy; culture and community; land and nature; sustainable water; local and sustainable food; materials and products; travel and transport; zero waste; zero carbon. https://www.bioregional.com

6 For a powerful and entertaining illustration of the failings of the planned economy, try Francis Spufford's novel *Red Plenty: Inside the Fifties Soviet Dream* (Faber & Faber, 2011).

7 Tom Crompton (2010) *Common Cause: The Case for Working with Our Cultural Values*. https://assets.wwf.org.uk/downloads/common_cause_report.pdf

8 See, for example, this piece by Max Lawson, Head of Advocacy and Public Policy, Oxfam Great Britain, on the World Economic Forum website: http://tinyurl.com/gsmfx6x. See also Wikipedia's posts on supply side economics and trickledown economics.

9 Credit Suisse 2017 Global Wealth Report (https://tinyurl.com/globalhwealth). By wealth, we mean the sum of all assets; house, money, pension fund, clothes, toothbrush – the lot.

10 See endnote 9 above.

11 Using data compiled from Giving What We Can (https://tinyurl.com/meanmedianwealth) as well as Credit Suisse. The factor of four is a rough but probably conservative estimate, based on guestimating the income distribution within the poorest half of Africa's population.

12 See Wilkinson and Pickett's *The Spirit Level: Why More Equal Societies Almost Always Do Better*. Also relevant is Andrew Sayer's book *Why We Can't Afford the Rich* (Policy Press, 2014). The Equality Trust website gives us a broad overview of the issues. www.equalitytrust.org.uk/

13 The Maxwell–Boltzmann distribution as defined for molecules moving in a gas between speeds c_1 and c_2:

$$f(c) = 4\pi c^2 \left(\frac{m}{2\pi k_B T}\right)^{3/2} e^{\frac{-mc^2}{2k_B T}}$$

- m is the mass of the molecule;
- k_B is the Boltzmann constant;
- T is absolute temperature.

14 Here is where the money spent on the UK National Lottery ends up:

- 53% is paid out in prizes;
- 25% goes to 'good causes' (some of which, some people argue, are causes that are rarely used by the people who buy most of the lottery tickets);
- 12% goes to the government as tax;
- 5% goes to the gambling company, Camelot, that runs it;
- 4% goes to the shop that sold the ticket.

So when you buy a ticket, you give away nearly half the money. And of that, just over half is to the 'good causes'. The tax you pay is particularly regressive since most lottery tickets are bought by people who have less available cash in the first place. The stats are from the Gambling Commission website. https://tinyurl.com/Nationalloterycon

15 KPMG International estimated that the Treasury will lose £1.1 billion over three years after the £2 limit comes into force (see https://tinyurl.com/UKgambling1). So that's roughly £350 million per year. FOBTs brought in £400 million in tax in 2017,and that's under a tax rate of 25% (https://tinyurl.com/UKgambling2). If only £50 million in tax is being paid at 25%, the annual revenue will have dropped from £1.8 billion to a relatively measly £200 million, resulting in a massive saving for punters.
16 Although problem gambling was estimated by GambleAware/IPPR to cost the economy £1.2 billion (which is well below the £2.6 billion the government collects in tax across the industry), this study did not take into account indirect effects on employers and financial burden and stress placed upon families of sufferers. The true cost may well be grossly underestimated. See https://tinyurl.com/UKgambling3
17 Reported by *the Guardian* on 15th June 2018. https://tinyurl.com/FOBTdelay
18 IPCC (2018) Global Warming of 1.5 °C – Summary for Policy Makers. http://report.ipcc.ch/sr15/pdf/sr15_spm_final.pdf
19 Rebecca Willis, me, Rosie Watson and Mike Elm (2020), The case against new coal mines in the UK, Green Alliance, https://bit.ly/2IRPTKk. This report takes apart a series of show-stopping bogus claims made in the successful planning application for the Woodhouse Colliery near Whitehaven in Cumbria. I think it will leave you wondering how on earth anyone was able to make them with a straight face, how they got through planning and how inadequate the secretary of state must have been in declining appeals to call in (meaning reassess) the local planning decision when its flaws were clearly pointed out. The mine will dig out coking coal, as used in traditional steel manufacture, emissions from which are expected to total more than 400 million tonnes of carbon dioxide.
20 Divestinvest (https://www.divestinvest.org/) is an organisation that exists to help individuals and organisations do just that. It outlines five criteria under which it would be healthy to remain invested in a coal oil and gas company. I think these are sound

guidelines but to my knowledge no fossil fuel company passes the test. The criteria are:

- No funding, directly or indirectly, of any propaganda or lobbying that is at odds with the science on the climate emergency or subverts global action in line with that science.
- A clear strategy to transition from fossil fuel at a pace that is in line with the science.
- No investment in exploration of further fossil fuel reserves.
- Science-based emissions targets for the company itself.
- Remuneration within the business set up to incentivise all staff to act in accordance with the above criteria.

21 You must always seek advice from a qualified financial advisor when setting up a company pension scheme. After looking at the marketplace for an ethical pension in 2019, there were only two group company pension scheme providers we felt comfortable investing our own pension funds in, Royal London and Aviva. The Royal London Workplace Pension enables selection of Sustainable and Ethical funds. Later Aviva UK launched the 'Stewardship lifestyle strategy' – a workplace pension default investment strategy that incorporates ethical and ESG (environmental, social and governance) considerations, which companies could select as their default.

22 Much of Lew's work was material drawn from the work of Deci and Ryan on self determination theory. For books, try, *Self Determination Theory: Basic Psychological Needs in Human Motivation, Development and Wellness* by Edward Deci and Richard Ryan (Guilford Press, 2017). Or for the slim line version, I recommend *Why We Do What We Do* by Edward Deci (Penguins Books, 1996).

Another model that I have found useful over the years, at least when asking what it will take for different groups to engage with climate change, is to consider it through the lens of three core psychological needs.

(1) **Relatedness** (belonging) – everyone likes to be part of a group. Few people like to be the only one acting sustainably. In this vein I was once told by a researcher who had been running focus groups on attitudes to climate change that if you wanted to engage people on the subject the three things you had to make them believe were (a) it would make them more like

everyone else, (b) it would make them more sexually attractive and (c) they would be more likely to get on TV.

(2) **Autonomy** (choice). People like to feel that they are in control of how they live. No one likes to be told by me or anyone else what they have to do to save the planet.

(3) **Competence.** The idea that we humans are basically incompetent at the moment is a hard message to take on board. To be told that how you live is not fit for the twenty-first century is not what any of us can easily hold on to unless we at least have some idea of what we might do differently.

23 Gini coefficient data from The Standardized World Income Inequality Database (SWIID). https://fsolt.org/swiid/. For a stack of interesting stats and charts on income inequality, I recommend browsing Our World in Data. https://ourworldindata.org/income-inequality

24 WeAll, an interesting start-up organisation focussed on the Wellbeing Economy. https://wellbeingeconomy.org/

'The purpose of the economy should be to achieve sustainable wellbeing – the wellbeing of all humans and the planet.

To achieve this goal, we need a major transformation of our world view, society and economy to:

1. Stay within planetary biophysical boundaries – a sustainable size of the economy within our ecological life support system.
2. Meet all fundamental human needs, including food, shelter, dignity, respect, gender equality, education, health, security, voice, and purpose, among others.
3. Create and maintain a fair distribution of resources, income, and wealth – within and between nations, current and future generations of humans and other species.
4. An efficient allocation of resources, including common natural and social capital assets, to allow inclusive prosperity, human development and flourishing. We all recognise that happiness, meaning, and thriving depend on far more than material consumption.'

25 BT's procurement guidelines in regards to climate change can be found here: https://groupextranet.bt.com/selling2bt/working/climateChange/default.html. They also outline their environmental principles on their website.

6 People and Work

1 See, for example, the late Swedish statistician Hans Rosling's entertaining and striking TED talks on population, health and wealth trends. Highly recommended, if you haven't seen them already. Very sadly he died in 2017. https://tinyurl.com/roslinghans

2 Stewart Wallis, former head of the New Economics Foundation, and before that International Director of Oxfam, estimates that this alone can cut the fertility rate by a massive 60%, making it, in his view, one of the world's most critical investments on three simultaneous fronts: morally, socially and environmentally. Stewart also writes: 'A study published in 2014 by Wolfgang Lutz, Director of the Vienna Institute of Demography, highlights why women's education is so important. In Ghana, for example, women without education have an average of 5.7 children, while women with secondary education have 3.2 and women with tertiary education, only 1.5.' See W. Lutz (2014) A population policy rationale for the twenty-first century. *Population and Development Review* 30(3), pp. 527–544.

3 This list is adapted from one provided by Stewart Wallis (thanks again Stewart) in turn drawing from Dr Malcolm Potts, University of Berkeley, School of Human Health.

4 See endnote 22 for Chapter 5 for Deci and Ryan's work. Also relevant is the concept of 'human givens' as drivers of human nature: defined as our psychological needs and thinking skills that compel us to act in particular ways. When our needs or expectations are not met we may experience stress, anxiety or anger. The approach behind human givens dictates that we are always looking to fulfil emotional needs (including security, attention, autonomy, emotional intimacy and *meaning and purpose*) using our 'innate resources' – emotional and social skills, which include imagination, empathy, memory and rationality. The Human Givens Institute concludes that emotional fulfilment allows us to function better in society, leading to better mental and

physical wellbeing. In order to reach emotional fulfilment, human psychology makes us want to *give*. See the book *Human Givens* by Joe Griffin and Ivan Tyrell.

5 All US prison stats are from the Bureau of Justice Statistics report: 'Correctional Populations in the United States 2015, NCJ 250374'.

6 All UK prison statistics are from the National Offender Management Service Annual Offender Equalities Annual Report 2016, available at www.gov.uk/statistics

7 Why are Norway's Prisons So Successful?, UK Business Insider. https://tinyurl.com/Norwayprisonstats

8 US National Institute for Justice. https://tinyurl.com/USrecidivism

9 Full Fact, Prisons: re-offending, costs and conditions, 2016. https://fullfact.org/crime/state-prisons-England-Wales/

10 See three references above.

11 Based on £4.26 billion per year cost of the UK prison service 2016–2017 (Statistica website: https://tinyurl.com/statisticaUKprisoncosts) and a prison population of 83,620 at the time of writing (www.gov.uk/government/statistics/prison-population-figures-2018).

12 The Economics of the American Prison System, Smartasset, 2018 (https://tinyurl.com/USprisoncosts) and *New York Times*, 23rd August 2013, City's Annual Per Inmate Cost is £168,000, study finds (referring to New York; see https://tinyurl.com/y7rje398).

7 Business and Technology

1 Bioregional: www.bioregional.com

2 The One Planet Living toolkit, see www.oneplanet.com. You have to sign up, but it is quick and painless.

3 The Science Based Targets initiative (SBTi) is a collaboration between CDP, World Resources Institute (WRI), the World Wide Fund for Nature (WWF) and the United Nations Global Compact (UNGC) and one of the We Mean Business Coalition commitments;

see http://sciencebasedtargets.org/#. So far over 400 have committed to SBTs, including over 100 global corporations.

4 Official Microsoft Blog, January 2020: Microsoft will be carbon negative by 2030. https://tinyurl.com/microsoftcarbontarget

5 'AI will be either the best, or the worst thing, ever to happen to humanity.' Stephen Hawking, speaking at the opening of the Leverhulme Centre for the Future of Intelligence, 2016.

8 Values, Truth and Trust

1 For the whole section on values I have drawn strongly on a commissioned paper for the Bellagio Initiative, 'The future of Philanthropy and development in the pursuit of human wellbeing', Tim Kasser, 2011: https://tinyurl.com/y7vuht95. Also, 'Common Cause: The Case for Working with our Cultural Values', Tom Crompton, 2010: https://assets.wwf.org.uk/downloads/common_cause_report.pdf

2 Carl Rogers, father of the person-centred approach to counselling. For example, his landmark 1954 book *On Becoming a Person: A Therapist's View on Psychotherapy.* The three core conditions are those required for a therapist to be helpful to their client. They apply equally well as the core conditions under which people can live well together.

Empathy: the ability to see the world from each other's point of view, understand and care about what that feels like.

Genuineness: a grounded honesty about what we are thinking and feeling and how we see the world.

Unconditional positive regard: belief in the inherent value of a person that is totally independent of anything at all that they may have done (murder, child abuse, terrorism, the lot).

3 Thanks to David Brazier for walking me through this stuff 20+ years ago, as well as the fascinating linkage he found between Carl Rogers' three core conditions for the therapeutic relationship and the antidotes to suffering as outlined in Zen Buddhist philosophy. This is now articulated in his book *Zen Therapy* (Constable & Robinson, 2001).

9 Thinking Skills for Today's World

1 In part what I'm saying here, as you may have spotted, is 'Please don't shoot *me* down too hard for *my* shortcomings' – and I'll try to treat you likewise!

2 The second edition of Jonathan Rowson's *Spiritualise* (available here: https://tinyurl.com/spiritualise) has a fuller articulation of the need for a multi-angled perspective and Perspectiva (www.systems-souls-society.com/), the organisation he co-founded with Tomas Bjorkman in 2015, is an interesting experiment along these lines.

3 A phrase coined by Tom Compton in his 2010 report for WWF and others 'Common Cause: the Case for Working with our Cultural Values'. Well worth a read. You don't solve bigger-than-self problems by appealing to individualist motivations. So you can't deal with climate change just by setting things up so that cutting carbon saves money. In fact you might make things worse by enforcing the individualist value that money might be the thing that matters most (https://tinyurl.com/68qrwdo). Sorry Tom if that's too crude a summary. I've missed plenty of nuance, and there was a reason why WWF and others commissioned a report, not a paragraph.

4 Thanks to David Brazier for many discussions in the 1990s on Zen and Tibetan Buddhism, and for his accessible and engaging book *Zen Therapy* (Constable & Robinson, 2001) which, among other things, drew out the parallels between Zen philosophy and practice, and the ideas of psychotherapist Carl Rogers. In particular, in Zen philosophy, the three core causes of suffering loosely translate into greed, hate and delusion. These are expressions of a person's loss of sight of the interdependence of all existence, and specifically the tendency to treat oneself as an independent island. When this happens, we get greedy for the things that seem good for that island, hateful of the things that don't seem to be and generally mixed up and deluded in our outlook. Each of the roots of suffering has an antidote, and these happen to map neatly onto what Carl Rogers described as the three

core conditions for the therapeutic relationship – the necessary conditions for the therapist to be of any help. Rogers' three core conditions, and the three Zen antidotes to suffering, are **empathy**, **genuineness** and **unconditional positive regard**. The antidote for greed is unconditional positive regard; if a person suffers from greed what they most need is to be accepted for who they are and however ugly their greed. If a person is angry (hateful) what they most need is empathy; to be understood. And if a person is deluded, what they need is your groundedness in reality and your truthfulness. Extremely interesting and not without powerful implications for challenges covered in this book. I'd write more but this is only supposed to be an endnote. Note finally that Rogers' core conditions feature strongly in the eight thinking skills that I have put forward.

Appendix: Climate Emergency Basics

1 Mike Berners-Lee and Duncan Clark, *The Burning Question* (Profile Books, 2012). Most of what we wrote then is now much more widely accepted, especially the need to leave the fuel in the ground. The numbers have moved on significantly in the seven or so years since then, as we have burned a good chunk of our remaining fuel budget and given ourselves a more urgent situation to deal with.

2 IPCC, 'Global Warming of 1.5 °C', October 2018. http://www.ipcc.ch/report/sr15/

3 W. Steffen, J. Rockström, K. Richardson et al. (2018) Trajectories of the Earth System in the Anthropocene. *Proceedings of the National Academy of Sciences* **115**(33), pp. 8252–8259. www.pnas.org/cgi/doi/10.1073/pnas.1810141115

The five feedback mechanisms were:

(1) Permafrost in the Arctic thawing and releasing methane, a powerful greenhouse gas which would trigger further warming and further melting of permafrost.

(2) Weakening of the capacity of land and sea to act as carbon sink that reduces atmospheric emissions and mitigates climate

impacts of emissions. Included here, for example, are forest fires from dried out woodlands in California, Portugal and elsewhere.

(3) Increasing bacteria in the oceans producing more carbon dioxide through respiration.

(4) Amazon forest dying back.

(5) Boreal forest dying back.

4 Data from Our World in Data. https://github.com/owid/co2-data

5 On closer inspection, the deviations from the pure exponential trend line are not just random noise. They mainly take the form of smooth wiggle; a sine wave in fact. There was a dip for the two world wars and the Great Depression, followed by a rapid growth spurt in the post-war years of rebuilding and oil discovery. After that came an oil crisis, followed by another bounce back above the trend line. Any maths and stats nerds might be interested to note that the wiggle turns out to be a lot like a sine wave with a period of 42 years. So the really interesting question is, was the sine wave caused by the wars, depressions, oil discoveries and so on, or were these events simply the means by which a more fundamental dynamic of the sine wave ripple on the exponential curve expressed itself on the ground? And if the latter, what brings that sine wave about, and what else is it connected to? (Thanks to Andy Jarvis at Lancaster University for pointing this out to me.)

6 The six-year rolling average is still above the trend line.

7 R. J. Millar, J. S. Fuglestvedt, P. Friedlingstein et al. (2017) Emission budgets and pathways consistent with limiting warming to 1.5 °C. *Nature Geoscience* **10**, pp. 741–747. I've taken the findings and extrapolated cumulative emissions forward from the end of 2014 to the end of 2017. Early in 2018, a letter in *Nature Geophysics* (A. P. Schurer, K. Cowtan, E. Hawkins et al. (2018) Interpretations of the Paris climate target, *Nature Geophysics* **11**, pp. 220–221) suggested an alternative way of feeding historical temperature records into the models than the way Millar et al. had used and this, fed into the method I've outlined above, would result in bringing forward the 1.5 °C overshoot date by around five years.

Carbon Brief has a nice discussion of the various estimates of the 1.5 °C budget. Unsurprisingly, the uncertainty stands out like a sore thumb. https://tinyurl.com/ybfcv7c8

8 See Millar et al. (2017) as in note 7 above. Supplementary information Figure S2 shows a roughly straight line fit of cumulative CO_2 emissions against temperature change with 2 °C for 1,200 Gt C, or 1 °C for every 2,200 Gt CO_2. The estimated temperature rise per tonne of carbon has been revised down by about 20% from the previous modelling, which led to the coining of the 'trillion tonne' limit – 1 trillion tonnes of carbon (3.67 trillion tonnes CO_2) for a 66% chance of less than a 2 °C temperature rise. From this paper you can infer that we are on track to overshoot the budget for a 1.5 °C temperature rise by about 2035 or, with very aggressive action on the non-CO_2 greenhouse gases, perhaps 2040.

9 This mine off the west coast of Cumbria is estimated to dig up coal with a carbon footprint of 8.4 million tonnes per year for 50 years. It has so far managed to wriggle through a series of cracks and loopholes in UK policy and planning guidance on the basis of being special coking coal supposedly (but not really) needed by the steel industry, on the basis that most will be exported and therefore not the UK's problem, and on the bogus basis that the rest of the world's coal production will somehow reduce in step with the production from this mine. For a detailed account of why this mine is such a dreadful idea see Rebecca Willis, Mike Berners-Lee, Rosie Watson and Mike Elm (2020), The case against new coal mines in the UK, Green Alliance; https://tinyurl.com/crazymine.

10 *SMARTer2030: ICT Solutions for 21st Century Challenges.* This 2015 report by the Global e-Sustainability Initiative (https://tinyurl.com/smarter2030) leaves rebounds out of its analysis except in the appendix and in one clear comment '*. . . policymakers need to provide the right conditions to ensure that emissions savings from ICT innovation do not lead to rebound effects within the macro-economy, as has been the case in the past*'. The treatment of rebounds in the appendix deals, I think, only with the most obvious rebound effects. A much more comprehensive analysis of rebound can be found in the UK

Energy Research Council's The Rebound Effect Report in 2007 (https://tinyurl.com/UKERCrebound) although even this misses off some of the widest long-term consequences.

11 Against this it is possible to argue that without the efficiency gains the growth curves would have been much steeper. However, a more credible argument is that the efficiency enables growth in utility at a slightly higher rate than the efficiency improvement. Therefore the net effect is an increase in the demand on resources.

12 Again, I've taken this from Millar et al. (2017); see note 7 above.

13 Carbon Engineering, 2020, Our Technology. https://tinyurl.com/sztu7br

Alphabetical Quick Tour

1 Duncan Clark and I used the following example in *The Burning Question*. If you buy a new car that does twice as many miles to the gallon the carbon savings get lost through the following mechanisms: (1) you drive a bit further, feeling less guilty and because it is cheaper, (2) you spend the money saved on something else that has a carbon footprint, (3) the oil supply chain adjusts its pricing and marketing, to increase sales to others and (4) the ease of extra mileage enables you to live further from the city centre, in a bigger house with bigger energy demands. It also leads to more road maintenance carbon costs.

2 See https://tinyurl.com/RSPBfarmbirds and https://tinyurl.com/GermanInsects

3 According to research from the University of Glasgow, looking at the health benefits of cycling to work. C. A. Celis-Morales, D. M. Lyall, P. Welsh et al. (2017) Association between active commuting and incident cardiovascular disease, cancer, and mortality: prospective cohort study. *BMJ* **357**: j1456, doi: https://doi.org/10.1136/bmj.j1456

4 S. H. Thilsted, A. Thorne-Lyman, P. Webb et al. (2016) Sustaining healthy diets: the role of capture fisheries and aquaculture for improving nutrition in the post-2015 era. *Food Policy* **61**, pp. 126–131.

5 Creating a Sustainable Food Future, Installment 8, from the World Resources Institute, covers water management in more detail. https://tinyurl.com/globalriceGHG

6 The fisheries expert Paul Danley and George Monbiot on conversation of shifting baselines. **https://tinyurl.com/shiftingbaselines**. The frog analogy is from Charles Handy, *The Age of Unreason* (Harvard Business School Press, 1989).

7 Death stats from Our World in Data. https://ourworldindata.org/causes-of-death#causes-of-death-over-the-long-run

8 Thanks to Roberto Muehlenkamp's entry in Quora which puts twentieth century deaths from war and oppression at 3.7%. I've added a bit for homicide. www.quora.com/What-was-is-the-most-violent-century-recorded-in-history

Notes on Units

1 Based on an Olympic swimming pool containing 2,500,000 kg of water, starting at 25 °C, the specific heat capacity being 4,200 kJ/kg and latent head of vaporisation being 2,265 kJ/kg.

2 Crude because all the different gases contribute to the greenhouse effect with different intensities and also have different half-lifes. We use the approximation that 1 kg methane has a global warming potential of 25 kg CO_2e over a 100-year period. But since methane is powerful but short lived, whereas carbon dioxide more or less lasts forever, if you take a 50-year timeframe, its global warming potential is more like 50 kg CO_2e. A better approximation still is to say that saving one tonne of methane per year forever has about the same effect as saving 2,700 tonnes of carbon dioxide.

3 Much more on this, of course, in my first book, *How Bad Are Bananas? The Carbon Footprint of Everything.*

INDEX

Locators in **bold** refer to tables; those in *italic* to figures

acidification of the oceans 62, 261
Africa
energy 67, 79
food wastage 44
inequality/wealth distribution 150–151
population growth 170
agriculture *see* food and agriculture
air pollution 123–125, 247
air travel 5, 242
food miles 34–36
impact of virtual meetings 129–130
low carbon 127–128
personal actions 129
risks of further growth 140
algal biofuels 88–89
Anderson, Kevin 129
animal sources of food 18–22, 259
inefficiency of animal feeds 242–243
laboratory grown meat 51–52, 75–77
micro-nutrients 20–21
protein 18–20, *19*
risks of further growth 138
Anthropocene 2, 69, 222–224, 243
antibiotics resistance 21–22, 29
appearances, over-valuing 205
appreciation, simple pleasures 140–141, 209–210, 213
armaments industry, and employment 172–173
artificial intelligence 189
atomic particles analogy of wealth distribution 154–158
Australia 77–79, *78*, 99–100
autonomous cars 126
autonomy/being in control 295
awareness *see* appreciation; self-awareness

balloon squeezing effect *see* rebound effects
Bangladesh *78*, 79–80, 240
batteries, storage of renewable energy 81
Belgium *78*, 79–80
belief systems 214–216, 265–266
belonging 294–295
Berners-Lee, Mike
Burning Question (with Clark) 4–5, 104, 244–245
How Bad Are Bananas? 36, 168, 256
big-picture perspective 208, 213, 222–224
biochar 104
bioenergy with carbon capture and storage (BECCS) 103
biodiversity 49, 60–62, 116–119, *117–119*, *119*, **119**, 243–244
big-picture perspective 222–223
pressure on land 88–89, 103
Bioregional, One Planet Living 181–183, *183*
boats/shipping 130–132, 264
Brazil 77–79, *78*

Brexit 244
Buddhism 215, 237–238
bullshit 201, 244; *see also* fake news; truth
Burning Question (Berners-Lee and Clark) 4–5, 104, 244–245
business as usual 9, 146, 233
businesses 179, 245
  environmental strategies 184–185
  fossil fuel companies 253
  perspectives/vision 180
  role in wealth distribution 157
  science-based targets 185–187
  systems approaches 180–183, *182–183*
  technological changes 188–190
  useful/beneficial organisations 179–180
  values 180, 196; *see also* food retailers

call centres, negative effect of performance metrics 143
calorific needs 13, 270–271
carbohydrates, carbon footprint 26–29, *28*
carbon budgets 58–59, 98, 166, 191–192, 231, 233–234
carbon capture and storage (CCS) 102–103, 159–160, 240–241, 245
carbon dioxide emissions, exponential growth 231–233, *232*, 250; *see also* greenhouse gas emissions
carbon footprints
  agriculture *25*, 25–29, 33–34
  carbohydrates *28*
  local food/food miles 34–36
  population growth 169
  protein *27*
  sea travel 130–132
  vegetarianism/veganism 30–31
carbon pricing 166–167, 239
carbon scrubbing 240–241, 245–246
carbon taxes 160, 163
CCS *see* carbon capture and storage
celebrities 205
change, embracing *see* open-mindedness
chicken farms 29
Chilean seabass (Patagonian toothfish) 37–38
China 246
  global distribution of fossil fuel reserves 99–100
  sunlight/radiant energy 77–79, *78*
choice//being in control 295
cities, urban planning and transport 119–122
citizen's wages 154–158, 174–175
Clark, Duncan: *Burning Question* (with Berners-Lee) 4–5, 104, 244–245
climate change (emergency) 3–5, 58, 62–63, 246
  big-picture perspective 222
  biodiversity impacts 60–62
  evidence against using fossil fuels 72–74
  ocean acidification 62
  plastics production/pollution 63–65, *63–64*
  rebound effects 59, 145–146, 186–187, 235–237, *236*
  science-based targets 185–187
  scientific facts 58–60, 229–241, *232*, *236*
  systems approaches 180–183, *182*
  values 191–192
coal 246; *see also* fossil fuels

comfort breaks, performance metrics 143–145
Common Cause report (Crompton) 147
community service 196
commuting 246; *see also* travel and transport
companies *see* businesses
competence 295
complexity 211, 213, 251; *see also* simplistic thinking
consumption/consumerism 247
  ethical 168, 190
  personal actions 197
  risks of further growth 139
  values 195
coronavirus disease (COVID-19) xvii
corporate responsibility 249; *see also* businesses
critical realism 198
critical thinking skills 210–211, 213
Crompton, Tom (*Common Cause* report) 147
cruises 132
cultural norms
  big-picture perspective 224
  values 193–194
cultures of truth 200–201
cumulative carbon budgets 58, 231
cycling 5, 114–116, 118, 132–133, 247

dairy industry 259; *see also* animal sources of food
democracy 159–160, 247–248, 268–269; *see also* voting
denial 225, 256
Denmark, wealth distribution *147–153*
Desai, Pooran 182–183
desalination plants, energy use 109
determinism 110–111, 248
developed countries 248
  energy use 109
  food waste 14, 45, 269
diesel vehicles 123–125, *125*
diet, sustainable 248; *see also* vegetarianism/veganism
digital information storage, and energy efficiency 95
direct air capture, carbon dioxide 103, 240–241, 245–246
distance, units of 271
double-sided photocopying metaphor 249
driverless cars 126

e-transport
  e-bikes 118, 132–133
  e-boats 131–132
  e-cars 118, 122, 249–250
  e-planes 127–128
  investment 159–160
economic growth 136, 249
  big-picture perspective 223–224
  carbon pricing 166–167
  carbon taxes 160, 163
  consumer power through spending practices 168
  GDP as inadequate metric 141–142, 144–145
  investment 159–160
  market forces 145–147
  need for new metric of healthy growth 142–145
  risks and benefits of growth 137–141, *138*
  trickledown of wealth 147–149, *148*
  wealth distribution *147–153*, 149–158
education 196, 249

efficiency 249
  digital information storage 95
  energy use 92–95
  investment 159–160
  limitations of electricity *82–97*, 95–97
  meat eating/animal feed 242–243
  rebound effects 94, 237
electric vehicles *see* e-transport
electricity, limitations of use *82–97*, 95–97; *see also* renewable energy sources
empathy 194, 208–209, 213
employment *see* work/employment
enablement, businesses 184–185
energy in a gas analogy of wealth distribution 154–158
energy use 66, 98, 111–112
  current usage 66–67
  efficiency 92–95
  fracking 89–91, *91*
  growth rates over time *see below*
  inequality 67, 100–102, 149
  interstellar travel 133–135
  limitations of electricity *82–97*, 95–97
  limits to growth 75–77, *76*, 110–111, 237–238
  nuclear fission 85–87
  nuclear fusion 87
  personal actions and effects 112–113
  risks of further growth 138
  sources 70–72
  supplied by food 13
  UK energy by end use 69–70, *70*
  units of 270–271
  values 191–192; *see also* fossil fuels; renewable energy sources
energy use growth 1–2, 67–69, *68*, 250
  and energy efficiency 94
  future estimates 108–110
  limits to growth 75–77, *76*, 110–111
  and renewables 91–92
enhanced rock weathering 104
enoughness 250; *see also* limits to growth
environmental strategies, businesses 184–185
  science-based targets 185–187
ethical consumerism 168, 190
ethics *see* values
evolutionary rebalancing 6, 250
expert opinion 251
exponential growth 137, *138*, 169, 231–233, 250–251
Extinction Rebellion (XR) xviii
extrinsic motivation and values 163–164, 192–193, 195

facts 251
  climate change 58–60, 229–241, *232*, *236*
  meaning of 197–198
  media roles in promoting 201–202; *see also* misinformation; truth
fake news 192, 197, 251; *see also* misinformation
farming *see* food and agriculture
fast food 266
feedback mechanisms 300–301; *see also* rebound effects
fish farming 37
fishing industry 36–40, 252
flat lining blip, carbon dioxide emissions 232–233, 250
flexibility *see* open-mindedness
flying *see* air travel

food and agriculture 12, 56–57, 252
animal farming 18–22, 32–33
biofuels 49
carbon footprints *25*, 25–29, *28*, 30–31
chicken farming 29
employment in agriculture 50, 252
feeding growing populations 52–53
fish 36–40
global surplus in comparison to needs 13, *14*
human calorific needs 13
investment in sustainability 55–56, 159–160
malnutrition and inequalities of distribution 16–17
overeating/obesity 17
personal actions 34, 38–39, 45, 49, 56–57
research needs 55
rice farming 33–34
soya bean farming *24*, 24
supply chains 54–55
technology in agriculture 50–52
vegetarianism/veganism 30–33; *see also* waste food
food imports, and population growth 170
food markets 147–149
food miles 34–36, 258–259
food retailers
fish 39–40
food wastage 45–47
rice 34
vegetarianism/veganism 31–32
fossil fuel companies 253
fossil fuels 70–72, 246, 252–253
carbon pricing 166–167, 239
carbon taxes 160, 163
evidence against using 72–74
global deals 97–102, *182*, 235, 238
global distribution of reserves 99–100, *101*
limitations of using electricity instead *82–97*, 95–97
need to leave in the ground 97–102, *182*, 235, 238, 252–253
sea travel 131
using renewables instead of or as well as 91–92
fracking 89–91, *91*, 253
free markets 145–147, 194, 257
free will 110–111, 189
frog in a pan of water analogy 264–265, 269
fun 253
fundamentalism 198, 215
future scenarios
aims and visions 9–10
climate change lag times 233–234
energy use 108–110
planning ahead 233–234
thinking/caring about 209, 213, 258
travel and transport 116–117, 126

gambling industry 157–158, 172–173, 292
gas analogy of wealth distribution 154–158
gas (natural gas) 253–254; *see also* fracking; methane
GDP
big-picture perspective 223–224
as inappropriate metric of healthy growth 141–142, 144–145
risks of further growth 139

genetic modification 51–52
genuineness 194
geo engineering solutions 254
Germany, tax system 164
Gini coefficient of income inequality *165*
global cultural norms 193–194, 224
global deals 184
  fossil fuels 97–102, 238, 240
  inequity 240
global distribution, fossil fuels 99–102, *101*
global distribution, solar energy 77–80, *78*, *101*
global distribution, wind energy 83, *84*
global food surplus 13, *14*
global governance 145–147, 159–160, 254–255
global solutions, big-picture perspective 223
global systems 5–7, 208, 254
global temperature increases 229–230
global thinking skills 208
global travel, by mode of transport *115*
global wealth distribution *147–153*, 150, 164, *165*
governmental roles
  big-picture perspective 223
  climate change policies 58–60, 229–241
  energy use policies 66, 113
  fishing industry 40
  promoting culture of truth 201–202
  sustainable farming 32–33, 50
  technological changes 190
  wealth distribution 156–157; *see also* global governance
greed 255; *see also* individualism
greenhouse gas emissions 238–239
  exponential growth curves 231–233, *232*, 250
  food and agriculture *25*
  market forces 145–146
  measurement 145
  mitigation of food waste 47–48, *48*, **48**
  risks of further growth 138
  scientific facts 58–60
  units 271–272; *see also* carbon dioxide; carbon footprints; methane; nitrogen dioxide
greenwash 245, 255
growth 255; *see also* economic growth; energy use growth; exponential growth

hair shirts 242, 253, 255–256
Handy, Charles 264–265
Happy Planet Index 143–144
Hardy, Lew 164
Hawking, Stephen 2–3, 188–189
Hong Kong, population growth 170
How Bad Are Bananas? (Berners-Lee) 36, 168, 256
hydrocarbons/hydrogen 81
hydroelectric power 83–85
hydro storage 81

ice 257
ICT (information and communication technology), impacts 95, 129–130
imperial units 270–272
income tax *see* tax system
India, global distribution of fossil fuel reserves 99–100
individual actions *see* personal actions and effects

individualism 136, 255, 257
indoor farming 51–52, 75–77
inequality 257
  and citizen's wage 174–175
  energy use 67, 100–102, 149
  food distribution 16–17
  global deals 240
  population growth 171
  prisons/prisoners 177
  tax system 162–166, *165*
  trickledown of wealth 147–149, *148*
  and values 191–193
  wealth distribution *147–153*, 149–158
insecurity 195
interdependencies, global/societal 211–212
Intergovernmental Panel on Climate Change 257–258
interstellar travel, impracticality of 133–135, 222, 265
interventionist economies 145–147
intrinsic motivation and values 163–164, 192–193, 195
investment 159–160, 257
  renewable energy sources 82, 98
  sustainable farming 55–56
iodine, malnutrition 16
IPCC *see* Intergovernmental Panel on Climate Change
Iraq, global distribution of fossil fuel reserves 99–100
Ireland, tax system 164
iron
  animal sources of food 20–21
  malnutrition and inequalities of distribution 16
irrigation technology 51–52
Italy, wealth distribution *147–153*, 151

Japan
  nuclear energy 86
  sunlight/radiant energy *78*, 79–80
Jevons paradox, energy efficiency 92–94
jobs *see* work/employment
joined up perspectives 211–214, 251
journalists *see* media roles

Kennedy, Bobby: speech on GNP 141–142
Keys to Performance (O'Connor) 203
kids 7–9, 209, 213, 258
kilocalories 13, 270–271
kinetic energy in a gas analogy 154–158

laboratory grown meat 51–52, 75–77
lag times, climate change 233–234
land requirements, sustainable travel 116–119, *117–119*, **119**
leadership 258
life expectancy, benefits of growth 140
life-minutes per person lost, diesel vehicles *125*
lifestyles 5; *see also* personal actions and effects
limits to growth 250
  big-picture perspective 222
  energy use 75–77, *76*, 110–111, 237–238
  twenty-first century thinking skills 209–210
  and values 192
local activities, appreciation of 140–141, 209–210, 213
local food, pros and cons 34–36, 258–259
luxury cruises 132

Maldives 240, 259
malnutrition 16–17
Marine Stewardship Council 37
market economies 145–147
materialistic values 196; *see also* consumption/consumerism
maturity, need for 109, *138*
Maxwell–Boltzmann distribution 154–156, 259, 292
measurement *see* metrics
meat eating *see* animal sources of food
media roles 259
  promoting culture of truth 201–202
  trust 205
messages, societal 195–196; *see also* values
methane 89–91, 238, 260
metric units 270–272
metrics
  healthy economic growth 142–145
  prisons/prisoners 178
  and values 196
  work/employment 171
micro-nutrients
  animal sources of food 20–21
  malnutrition 16
Microsoft, carbon pricing scheme 167
mindfulness 197, 213, 215
misinformation 251
  and trust 204, 206
  and truth 197
  and values 192
mitigation strategies, businesses 184–185
models, climate change 229–230, 233–234
molecular analogy of wealth distribution 154–158
Monbiot, George 264
motivation
  extrinsic/intrinsic 163–164, 195
  and trust 203, 206
Musk, Elon 189

natural gas 253–254; *see also* fracking; methane
neoliberalism 50, 147–149, 194, 257, 260; *see also* free market
Netherlands *78*, 79–80, 170
'net-zero' targets xix
neuroscience 260
nitrogen dioxide 124, 238
Norway *147–153*, 157, 176–177
nuclear fusion 87, 261
nuclear power (fusion) 85–87, 261

obesity 17
ocean acidification 62, 261
ocean planting 103
O'Connor, Tim: *Keys to Performance* 203
offsetting 105
oil 261; *see also* fossil fuels
One Planet principles 181–183, *183*
open-mindedness
  neuroscience 260
  respect for 202
  spirituality/belief systems 214
  and trust 204–206
optimism bias 261–262
over-simplification 205; *see also* complexity
overeating 17

pandemic xvii
parental responsibility 262
Paris climate agreement 187
particulate air pollution 123–125
Patagonian Toothfish 37–38

pay rates 195; *see also* wealth distribution
personal actions and effects 225–227, 262
air travel 129
antibiotics resistance 22
climate change 62–63
energy 112–113
feelings of insignificance in global systems 5–7
food/agricultural issues 34, 38–39, 45, 49, 56–57
population growth 170–171
promoting culture of truth 201
technological changes 190
values 197
wealth distribution 157–158
work/employment 173
'personal truths' 199
perspectives
big picture 208, 213, 222–224
businesses 180
joined up 211–214, 251
photocopying metaphor 249
photovoltaic technology 71, 74–75; *see also* solar energy
physical growth mindset 137
Planet B, lack of 133–135, 222, 265
planned economies 145–147
planning ahead, future scenarios 233–234
planning, urban 119–120
plastics 63–65, *63–64*, 262
politicians *see* governmental roles; voting
pollution, chicken farming 29; *see also* air pollution
population growth 169–170, 262–263
feeding growing populations 52–53
investment in control measures 159–160, 171
personal actions and effects 170–171
risks of further growth 139
positive feedback mechanisms, climate change 229–230, 267
power units of 270–271
prisons/prisoners 175–178, *177*, 196, 263
problem-solving methods 6
profit motive 180, 196
protein
animal sources 18–20, *19*
carbon footprints 26–29, *27*
protests xviii
psychology 256
public service 196

reader contributions 10–11
ready meals 266
rebalancing, evolutionary 6, 250
rebound effects 243, 263, 300–301
business strategies 184
climate change 59, 145–146, 186–187, 235–237, *236*
energy efficiency 94, 237
virtual meetings 129–130
reductionism 211–212, 215
refugees 263
relatedness/belonging 294–295
religion 214–216
renewable energy sources 72, 237–238, 263
hydroelectric power 83–85
investment 159–160
limitations relative to fossil fuels *82–97*, 95–97
using instead of/as well as fossil fuels 91–92
wind energy 82–83; *see also* biofuels; carbon capture and storage; solar energy
respect 193–194, 202, 224

responsibility
  corporate 249
  parents 262
  super-rich 152
restaurants role
  food wastage 45
  vegetarianism/veganism 32
retailing, food *see* food retailers
revenge, prisoners 176–177
rice farming 33–34, 51–52, 263–264
rock weathering, carbon capture and storage 104
Rogers, Carl 194
Russia 240, 264
  global distribution of fossil fuel reserves 99–100
  sunlight/radiant energy 77–79, *78*
Rwanda *78*, 79–80, 195

salaries 195; *see also* wealth distribution
Science Based Targets Initiative (SBTi) 185–187
scientific facts *see* facts
scientific fundamentalism 198
scientific reductionism 211–212, 215
seabass, rebadging Patagonian toothfish as 37–38
sea travel 130–132, 264
self-awareness
  of simple/small/local 140–141, 209–210, 213
  and trust 203–204, 206
self-reflection, twenty-first century thinking skills 210
sentient animals, treating decently 12, 18
shared-use vehicles 120–122
shareholder profits 180, 196
sharing 167
shifting baseline syndrome 264
shipping 130–132, 264
shock 264–265
simple things, appreciation of 140–141, 209–210, 213
simplistic thinking 205; *see also* complexity
slavery
  and citizen's wage 174
  and employment 171
  fishing industry 36, 38–39
slowing down 209–210, 223
small scale, appreciation of 140–141, 209–210, 213
Smith, Adam: *The Wealth of Nations* 147
social support structures, and values 196
soil carbon sequestration 103
solar energy 265
  amount falling on Earth 74
  coping with intermittent sunlight 80–82
  countries with highest radiant energy 77–80
  countries with least radiant energy 79–80
  relative to fossil fuel reserves *101*
  global distribution of radiant energy 77–80, *78*
  harnessing 74–75
South Korea, sunlight/radiant energy *78*, 79–80
soya beans *24*, 24, 265
space tourism 109, 116
spaceflight, impracticality of interstellar travel 133–135, 222, 265
Spain, wealth distribution *147–153*, 151
spending practices, ethical consumerism 168, 190

spirituality/belief systems 214–216, 265–266
status symbols 195
sticking plasters (band aids) 266
storage of renewable energy 80–82
sunlight *see* solar energy
supermarkets *see* food retailers
super-rich
  responsibilities 152
  taxation 165–166
  wealth distribution 155–156
supply chains
  ethical consumerism 168
  food and agriculture 54–55
  science-based targets 186–187
systems approaches
  big-picture perspective 223
  businesses 180–183, *182*
  One Planet Living principles 181–183, *183*

Taiwan, tax system 164
takeaways 266
tax system 266–267
  carbon taxes 160, 163
  wealth distribution 157, 162–166
technological changes 267
  agricultural 50–52
  big-picture perspective 222–223
  business strategies 188–190
  and economic growth 140
thinking skills
  big-picture perspective 224
  twenty-first century 207–214, *212–213*
Thunberg, Greta xviii
tipping points *see* trigger points
town planning 119–120
transmission of renewable energy 82
travel and transport 114
  air travel 126–130
  autonomous cars 126
  commuting 246
  current rates 114–116, *115*
  cycling 132–133
  diesel vehicles 123–125, *125*
  e-cars 122
  food miles 34–36
  future demands 116–117, 126
  land needed for sustainable 116–119, *117–119*, *119*, **119**
  sea travel 130–132
  shared-use vehicles 120–122
  spaceflight 133–135
  urban 119–122
tree planting 102
trickledown of wealth 147–149, *148*, 267
trigger points, step changes in climate 2, 229–231
trust 203–206
truth 197–198, 267
  big-picture perspective 224
  importance of seeking 199–200
  media roles 201–202
  'personal truths' 199
  promoting culture of 200–201
  respect for 193–194
  and trust 203–206
tsunami, December 2004 2
twenty-first century thinking skills 207–214, *212–213*, 224
2-degree 'safe limit' for temperature rise 59, 229–230, 233–234, 267

unconditional positive regard 194
United Kingdom
  energy by end use 69–70, *70*
  gambling industry 158
  nuclear energy 86
  population growth 170
  prisons/prisoners 176

United Kingdom (cont.)
- sunlight/radiant energy *78*, 79–80
- wealth distribution 154–155

United States
- global distribution of fossil fuel reserves 99–100
- prisons/prisoners 175–177
- sunlight/radiant energy 77–79, *78*
- tax system 164
- wealth distribution *147–153*, 149–152

units, metric/imperial 270–272
urban planning 119–120
urban transport 119–122

value of human life 268
values 7–9, 191
- big-picture perspective 224
- businesses 180, 196
- changing for the better 194–197
- and economics 136
- evidence base for values choices 191–193
- extrinsic/intrinsic 192–193, 195
- global cultural norms 193–194, 224
- prisons/prisoners 177
- technological changes 189–190
- wealth distribution 150
- work/employment 173; *see also* ethical consumerism

vegetarianism/veganism 30–33
Venezuela, global distribution of fossil fuels 99–100
violent deaths 268
virtual travel 129–130
visions of future 9–10
- businesses 180

vitamin A 16, 20–21, **275**
voting, power of 268–269
- climate change policies 58–60, 229–241
- energy policies 66, 113
- promoting culture of truth 201–202; *see also* democracy

waking up 269
Wallis, Stewart 165–166
waste food 40–49, 269
- mitigation 47–49, *48*, **48**
- as proportion of food grown 13–16, *15*
- by region/type/processing stage **41**, *43–44*, *44*

water use technology, in agriculture 51–52
watts 13, 270–271
wealth distribution
- economics *147–153*, 149–158
- tax system 157, 162–166, *165*; *see also* inequality

The Wealth of Nations (Smith) 147
weapons industry 172–173
weight, units of 272
wellbeing 269
- benefits of growth 140
- businesses, role of 179–180
- and citizen's wage 174
- metrics of healthy growth 143–144
- work/employment 171–173

Wellbeing Economy 295
wind energy 82–83
wisdom, need for 109, *138*
work/employment 258
- agricultural work 50, 252
- and citizen's wage 174–175
- investment in sustainability 56
- personal actions and effects 173
- useful/beneficial 171–173
- values 173

zinc 16, 20–21